T0313353

Manufacturing Cost Policy Deployment (MCPD) Transformation

Uncovering Hidden Reserves of Profitability

Manufacturing Cost Policy Deployment (MCPD) Transformation

Uncovering Hidden Reserves of Profitability

Alin Posteucă

CRC Press
Taylor & Francis Group
Boca Raton London New York

CRC Press is an imprint of the
Taylor & Francis Group, an **informa** business

A PRODUCTIVITY PRESS BOOK

CRC Press
Taylor & Francis Group
6000 Broken Sound Parkway NW, Suite 300
Boca Raton, FL 33487-2742

© 2018 by Taylor & Francis Group, LLC
CRC Press is an imprint of Taylor & Francis Group, an Informa business

No claim to original U.S. Government works

Printed on acid-free paper

International Standard Book Number-13: 978-1-138-09392-8 (Hardback)
International Standard Book Number-13: 978-1-315-10632-8 (eBook)

Visit the Taylor & Francis Web site at
http://www.taylorandfrancis.com

and the CRC Press Web site at
http://www.crcpress.com

Contents

List of Figures

List of Tables

List of Abbreviations

ABC	Activity-based costing
AMCIB	Annual manufacturing cash improvement budget
AMIB	Annual manufacturing improvement budgets
BPC	Best practice cost
CAL	Current admissible losses
CCILW	Costs of current inadmissible losses and waste
CCLW	Critical costs of losses and waste
CF	Continuous flow
CIL	Current inadmissible losses
CPM	Company productivity mission
CPS	Company productivity strategy
CPV	Company productivity vision
DCM	Daily cost management
DLT	Delivery lead time
DMAIC	Define, measure, analyze, improve, and control
DMIs	Daily management indicators
ECRS	Elimination, combination, rearrangement, and simplification
ELT	Equipment loading time
EWH	Equipment working hours
FLT	Factory lead time
IE	Industrial engineering
JIT	Just-in-time
KKIs	Kaizen and kaikaku indicators
KPIs	Key performance indicators
LM	Lean manufacturing
MACPD	Marketing and administrative cost policy deployment
MaLT	Material lead time
MB	Management branding

MBO	Management by objectives
MBR	Management by results
MCI	Manufacturing cost improvement
MCICP	Manufacturing cost improvement catchball process
MCImi	Manufacturing cost improvement means impact
MCKP	Manufacturing cost key point
MDC	Methods design concept
MKP	Manufacturing key points
MLT	Manufacturing lead time
MR	Means reference
MRT	Means-reduction target
MTT	Means target touch
OMIs	Overall management indicators
OPF	One-piece flow
OPL	One-point lesson
OTIF	On-time in-full
PBA	Price benchmark analysis
PBM	Productivity business model
PCBG	Productivity core business goals
PDCA	Plan–do–check–act
PFC	Product family cost
PL	Physical loss
PMP	Productivity master plan
PST	Problem-solving techniques
R&D CPD	R&D cost policy deployment
RCA	Root cause analysis
RMCA	Reverse manufacturing costing analysis
ROI	Return on investment
SC	Standard costing
SCCPD	Supply chain cost policy deployment
SCM	Supply chain management
SKPMP	Strategic key points on manufacturing processes
SMART	Specific, measurable, attainable, realistic, and timely
SOP	Standard operating procedure
SWOT	Strengths, weaknesses, opportunities, and threats
TCCLW	Total critical costs of losses and waste
TCLW	Total costs of losses and waste
TCPD	Total cost policy deployment
TLT	Total lead time

TLW	Total losses and waste
TPM	Total productive maintenance
TQM	Total quality management
TRL	Time-related loss
TTP	Time-to-production
TTSP	Time-to-start-production
VAOT	Value-adding operating time
WCM	World class manufacturing
ZCLW	Zero costs of losses and waste

Preface

Achieving a long-term acceptable level of manufacturing profitability through productivity requires the total commitment of management teams and all staff in any manufacturing company and beyond.

Achieving this acceptable level of productivity or multiannual manufacturing target profit can be accomplished by meeting the external target profit expected from sales and by achieving the internal target profit by continually improving the manufacturing costs. The role of productivity is decisive, both in meeting the external target profit, by providing the manufacturing capacities needed to support sales plans (effectiveness or maximizing output by reducing or eliminating the noneffective use of inputs—losses), and in achieving internal profit through continuous improvement of manufacturing costs (efficiency or minimization of inputs by reducing or eliminating excessive amounts of inputs—waste).

Awareness and continuous improvement of manufacturing costs behind losses and waste is the core goal of the Manufacturing Cost Policy Deployment (MCPD). Achieving this goal will continually uncover the hidden reserves of profitability through a harmonious transformation of the manufacturing flow, coordinated by the continuous need to improve manufacturing costs. Setting annual targets and means for manufacturing costs improvement, more exactly for costs of losses and waste, and the exact fulfillment of these, requires mobilization of all people in the company to carry out systematic improvement activities (kaizen) and systemic improvement actions (kaikaku) of the processes of each product family cost.

The MCPD system was born out of careful observation of the challenges, principles, and phenomena of manufacturing companies and the profound discussions with the people in these companies at all levels. At the same time, the associated theories available were analyzed.

This book is organized in three sections. The first section presents the concept and the need for an MCPD system from a managerial perspective. In the second section, the transformation of manufacturing companies through the MCPD system is presented, more precisely the details of the initial steps of the implementation of the MCPD, the three phases and the seven steps of the MCPD, and the elements necessary for a constant and consistent application of the MCPD. In the last section, there are two examples of the MCPD implementation in two different types of industries, namely, manufacturing and assembly industry and process industry, and two case studies for the improvement of manufacturing costs for each (cost of equipment setup loss, using kaizenshiro; replacement of bottleneck equipment and associated costs of losses, using kaikaku; cost of quality losses with improving operators' skills to sustain quality, using kaizen; and cost problem solving with the consumption of lubricants for one of the equipment, using A3).

The main audience of this book is made up of top managers, middle managers, and professional support staff who manage manufacturing areas. The continuous coordination of all manufacturing improvement through the need to improve manufacturing costs requires a close collaboration between engineering specialists and those in economics, especially in cost and managerial accounting.

Therefore, the author's hope is that the MCPD system will provide a way of addressing the difference between manufacturing targets profit and external targets profit expected from sales by continuously uncovering the hidden reserves of internal profit that are obtained directly from manufacturing processes by increasing productivity.

Acknowledgments

Over the years, I received many valuable ideas and comments from many people to reach the current form of MCPD. I am most grateful to John Heap, president of the World Confederation of Productivity Science, for his constant professional inspiration for more than ten years. I also thank Shigeyasu Sakamoto, president of Productivity Partner Incorporation, who has helped me better understand the Japanese way of achieving process improvements during our discussions.

I also thank Michael Sinocchi and Alexandria Gryder of Productivity Press for their professionalism and for their valuable efforts and advice on this book and more.

I thank Vijay Bose of SPi Global for his professionalism on content editing the manuscript and Jay Margolis of CRC Press for the careful co-ordination of the production of this book.

Since MCPD is the result of long synthesis and distillation, being successfully implemented, totally or partially, in several manufacturing companies, I thank all managers from different positions in these companies for giving me the chance to observe and discuss production phenomena and principles and all those who contributed directly or indirectly to the development of MCPD system.

Last but not least, I sincerely thank and dedicate this book to my family. In this regard, I thank my wife Emiliana for her total support and constantly inspired confidence to write this book. Full of hope and confidence, I thank my daughter Andreea and my son Ştefan for their understanding and for the many moments with them that added to my inspiration.

Alin Posteucă

About the Author

Alin Posteucă is a management consultant in profitability, productivity, and quality and managing partner of Exegens Management Consultants (Romania). Prior to this position, he held top management positions in manufacturing and service companies.

His recent research includes the development of the Manufacturing Cost Policy Deployment (MCPD) system for multiannual targets profit by identifying new manufacturing profit opportunities by exploiting manufacturing cost improvement through continued productivity growth. The main purpose of the MCPD system is to meet the annual manufacturing cost improvement goal. The annual manufacturing cost improvement goal is the difference between the annual manufacturing target profit and the annual external profit expected from sales, for each company in the group and for every product family cost of each company.

He has been actively involved in various industrial consulting and training projects for more than 20 years in Romania and has published in various research journals and presented papers at numerous conferences, regarding productivity, profitability, and quality.

He received his PhD in industrial engineering from the Polytechnic University of Bucharest (Romania) and his PhD in managerial accounting from the Bucharest University of Economic Studies (Romania). He received his MBA from the Alexandru Ioan Cuza University of Iaşi, Romania. Also, he is a certified public accountant in Romania.

About the Author

Introduction

The Manufacturing Cost Policy Deployment (MCPD) system is about achieving the multiannual manufacturing target profit by targeting strategic productivity improvements over the long, medium, and short term.

The multiannual manufacturing target profit is achieved through the multiannual external profit expected from sales and through the multiannual internal profit achieved through continuous productivity improvement. The multiannual external profit expected from sales is the usual way to plan and earn profit starting from the sales plan (connected to the productivity vision and mission—sales volumes, market share, and profit) and then through developing the production plan, the supply chain plan, the human resources plan, the inventory plan, and the administrative and funding plans. The multiannual internal profit is determined by the difference between the multiannual target profit and the multiannual external profit expected from sales, in order to ensure a reasonable level of profit for an acceptable development of a manufacturing company. At one year, the difference between the annual target profit and the annual external profit will be achieved from either the increase of the annual external profit from sales (by increasing the volume of production sold, implicitly the use of production capacities) or the reduction of annual expenses (through manufacturing cost improvement, MCI). The difference between the annual target profit and the annual external profit from sales already adjusted is called annual manufacturing cost improvement goal (annual MCI goal). The annual MCI goal is planned to be achieved for each company within the group and for every product family cost (PFC) of each company. The PFCs are those product groups that have about the same pressure for MCI, the same need for cost reduction and unit profit growth, and which have about the same opportunities to improve losses and waste as they run around the same manufacturing flow processes.

Synergy of the MCPD System at the Process Level

Once the annual MCI goal level at a PFC level, the top-down approach, or the annual demand for MCI is established, the annual MCI targets will be set for each manufacturing flow process (for each PFC, for existing and future products). The annual MCI targets will be set for the cost of losses and waste (CLW, the chronic cost problems) and for the critical cost of losses and waste (CCLW, the acute cost problems). CLW represents the transformation of losses and waste into costs through a scientific approach to unnecessary minutes and unnecessary material consumption. CCLW is the systemic approach to CLW, namely, the approach to CLWs that produce chain effects across the entire manufacturing flow and beyond. The annual CCLW sums up part of the annual CLW. The annual CLW and the annual CCLW represent the total offer for MCI. From this offer, one will use exactly what it takes to meet the annual MCI goal by setting annual MCI targets based on annual MCI means. Through the annual reconciliation between the demand for MCI (annual MCI goal) and the offer for MCI (CLW and CCLW), the *annual uncovered hidden reserves of profitability obtained by continuously transforming the manufacturing flow of each PFC according to the market signals* (price and profit) are determined. The annual reconciliation is based on the MCI catchball process through which, following several rounds of negotiations between managers and specialists at all levels of the company, an annual consensus on the level of MCI targets and means to meet the annual MCI goal is obtained. The true value of a manager is his ability to accurately plan the winning activities.

The strategic approach to productivity improvement is to establish annual MCI means to meet annual MCI targets at the process level of a PFC. Establishing annual targets and means for MCI for each PFC process represents the annual MCI policy deployment. The annual MCI means are defining both the strategic improvement project (systematic, kaizen, and systemic, kaikaku) and solving the daily problems of MCI targets. In establishing the annual MCI targets, the priority is given to PFCs with the highest external pressure on MCI, and in setting MCI means, the priority is given to CCLW targets (addressing the root causes to meet the annual MCI goal). The setting of annual MCI means targets is achieved by setting targets for losses and waste that are converging to meet the annual MCI targets.

The MCI targets and means (*Phase 1: Manufacturing Cost Policy Analysis; Step 1: Context and Purpose and Step 2: Targets and Means for MCI*) are achieved through the development and implementation of the annual manufacturing improvement budget and the annual action plan for MCI means (*Phase 2: Manufacturing Cost Policy Development; Step 3: Annual Budgets for MCI and Step 4: Action Plan for MCI*), engaging the workforce to achieve the MCI targets, MCI performance management, and daily MCI management (*Phase 3: Manufacturing Cost Policy Management; Step 5: Engage the Workforce for MCI, Step 6: MCI Performance, and Step 7: Daily Management*). All these together represent the MCPD system (three phases and seven steps).

In this context, the MCPD system addresses the top management that has the need to meet the multiannual and annual target profits through strategic and operational enhancement of productivity improvements.

How This Book Is Organized

This book is organized in three sections. The first section (Introduction to the MCPD System) describes the need for MCPD by presenting the connections between the need for continuous reduction of costs and MCI (Chapter 1) and presenting the MCPD system within the dynamics of business contexts (Chapter 2). The second section (MCPD Transformation) presents the establishment of an MCPD system and steps to begin (Chapter 3); the three phases and seven steps of the MCPD system implementation, by designing, building, and full development at the level of each PFC (Chapter 4); and how to apply constantly and consistently the MCPD system (Chapter 5). The third and final section (MCPD Practical Implementations and Case Studies) presents two transformations of the manufacturing flow through the MCPD system, aiming at achieving the annual MCI goal and the annual target profit, in two different types of industries, namely, manufacturing and assembly industry and process industry (Chapter 6).

The following is an overview of the six chapters in this book.

Chapter 1: Starting from the Need for Continuous Cost Reduction

This chapter begins by presenting the need for the MCPD system. Five main barriers for the consistent and harmonious transformation of manufacturing companies are presented. Then the MCPD system is defined, described, and positioned within the manufacturing companies to show the connection between the need for continuous cost reduction and

MCI. This connection is based on the seven principles of the MCPD system described in this chapter, along with the basic features of the MCPD and with the help of transposition of the cost strategy into action (top-down and bottom-up approaches of MCI). Further on, a synthetic example of implementing the annual MCI policy deployment is presented to understand how to set up the annual MCI targets and means. The chapter ends by presenting the stakes of the MCPD system: *uncovering hidden reserves of profitability.*

Chapter 2: MCPD System: Overview and Dynamics of Business Contexts

This chapter presents theoretically the MCPD system and its medium- and long-term approaches. The chapter begins by presenting the transition from productivity vision to annual action plans for each PFC. The benefits of the MCPD system are presented for every hierarchical level in the company. Then the goals and approaches of the three phases and the seven steps of the MCPD system are presented. From the perspective of the MCPD system, the main ingredients of long-term manufacturing target profit is presented, namely, price benchmark analysis and synchronization of life cycle targets (for profit, price, sales, capacity, productivity, and cost) and of efficiency of investment and productivity. Finally, how to strategically align the opportunities for improvement to the need for MCI is presented.

Chapter 3: Establishment of an MCPD System: Steps to Begin

What are the steps preceding the introduction of the MCPD system? This is the question that is being answered in this chapter. The answers are two way: a technical answer (MCI targets deployment: preliminary steps) and an organizational answer (preparation for the implementation of an MCPD system). Therefore, this chapter presents first the preliminary steps needed to establish MCI targets, by presenting in detail: the market driven activities for setting annual MCI goal, the profit-driven activities for setting annual MCI targets and the annual management coordinated by MCI targets and means deployment. Further on, the organizational steps necessary for the adoption of the MCPD system are presented, as a way of fulfilling the multiannual production profit plan by productivity improvement;

more precisely, the elements underlying the MCPD system implementation decision and the prediction of its effects, the MCPD system organization, and the creation of its structures to support MCI and internal and external communication of the MCPD system purpose are presented.

Chapter 4: MCPD Implementation: Designing, Building, and Full Development

This chapter focuses on the detailed presentation of the three phases and the seven steps of the MCPD system. In the first phase of the MCPD system, manufacturing cost policy analysis, the first two steps of MCPD are presented. In these two first steps, the annual MCI targets and means are set for each PFC through annual reconciliation (top-down and bottom-up for setting annual MCI goals and annual MCI targets and means). The second phase, manufacturing cost policy development, addresses the annual manufacturing improvement budgets development for existing and new products and the annual manufacturing cash improvement budget, in order to support the annual MCI targets and means (step 3) and the annual action plan for MCI means, including the annual individual plans for MCI (step 4). In the third phase of the MCPD system, manufacturing cost policy management, the workforce is engaged to achieve the MCI targets through departmental organization for achieving MCI targets, through the development of an annual MCI training plan by running of the improvement activities and actions of the annual MCI means to meet annual MCI targets (step 5); the performance level of the current state of MCI against annual MCI targets is followed by developing the annual MCI performance management (step 6); and by daily MCI management (step 7), the tangible and intangible effects of annual MCI are monitored daily and the deviations of MCI targets and contextual managerial behaviors are solved.

Chapter 5: MCPD Constant and Consistent Application

How is the MCPD supported on the long term at all levels of the organization? What is the impact of the MCPD system on the continuous transformation of the manufacturing flow for each PFC? These are the questions that have been answered in this chapter. In order to answer the first question, the following were presented: the MCPD system information centers and horizontal and vertical communication, continuously collecting and recording the data and information; management branding and managerial support for the MCPD system; and continuous monitoring of the annual MCI goal for each PFC (or the *annual uncovering of hidden reserves of profitability*). To answer the second question, the impact of the MCPD system on manufacturing lead time and beyond, on work in progress (WIP), and on material stock was presented.

Chapter 6: Applications of the MCPD System

This last chapter presents two applications of the MCPD system in two companies in two different industries, namely, manufacturing and assembly industry and process industry. The main actions, activities, and challenges for starting the MCPD implementation are presented; the MCPD steps and two case studies for each company (MCI means) (only a few of the annual and multiannual strategic projects of productivity improvement to meet the annual MCI goal and multiannual internal profit and target profit) are also presented. For the first company, MCI by improving the equipment setup and adjustment time and associated costs (with *kaizenshiro*) (case study 1) and MCI by increasing productivity with the replacement of bottleneck equipment (with *kaikaku*) (case study 2) are presented. For the second company, MCI by improving operators' skills to sustain quality (with *kaizen*) (case study 1) and daily MCI management—cost problem solving for lubricants consumption for one of the equipment—(with A3) (case study 2) are presented. The results of applying the MCPD system in the two companies by continuously targeting productivity improvements through the need to meet the annual MCI goal and multiannual internal profit and target profit have been fully achieved.

I am confident that the MCPD system will help your company to meet multiannual and annual profit plans by improving productivity.

INTRODUCTION TO THE MCPD SYSTEM

Chapter 1

Starting from the Need for Continuous Cost Reduction

The development of manufacturing companies will never cease. For this reason, they are constantly transforming or articulating/rearticulating according to the market needs through a continuous search and improvement of the best production method. Often, production companies are manufacturing increasingly complex products that contain many components originating from different industries. The realization of these products implies the use of a method for transforming resources (inputs of labor, equipment, materials and raw materials, information, spaces, and utilities) into results (outputs expected by customers, by employees, by suppliers, by shareholders, by the society, and by the community in general) so that manufacturing companies can ensure consistent development and long-term profitability. The most commonly used methods of transforming resources into expected results are *Total Quality Management* (TQM), focusing on meeting customer needs and organizational objectives; *Total Productive Maintenance* (TPM), focusing on maximizing equipment effectiveness (maintenance and equipment losses); *Industrial Engineering* (IE), focusing on maximizing effectiveness and efficiency of operators (workplace organization and waste); *Just-In-Time* (JIT), focusing on logistics especially to reduce the inventory levels and ensure timely deliveries (logistic and inventory); *Lean Manufacturing* (LM), focusing on creating value for customers (market, flow, leveling, and improvement); *Six Sigma*, focusing on quality improvement (quality and defects); and *World Class Manufacturing* (WCM), focusing on the unification of the basic concepts of TPM and LM with TQM. However, depending on the specifics of current and especially future work, many manufacturing companies use a mix of these methods and define their own production system using designations such as "X" Production System (X being the company name). The common denominator of these companies is the same: ensuring an acceptable level of effectiveness and efficiency of the production method (or an acceptable level of productivity). In this context of the need to ensure an acceptable level of productivity, the continuous reduction of the costs of all resources used is a constant desideratum for the management of many companies, especially when it comes to the rigorous and pertinent planning of unit costs reduction to ensure an acceptable level of competitiveness through price and long-term profit margin.

In this context, in Section 1.1 we address the need for the MCPD system, the five main barriers to the consistent and harmonious transformation of manufacturing companies and the presentation of the MCPD system (definition, description, positioning, principles, and basic characteristics). Section 1.2 shows the connection between cost reductions and MCI. Section 1.3 addresses how to translate the cost strategy into daily activities by continually reconciling the top-down approach with the bottom-up approach for the annual setting of MCI. Section 1.4 addresses the setting of MCI targets and means at the level of each product family cost (MCI policy deployment). Further, Section 1.5 provides a synthetic example of implementing the annual MCI policy deployment to understand how to set annual MCI targets and means to meet internal profit targets and support external target profit. This chapter ends by presenting the stakes of the MCPD system: *uncovering hidden reserves of profitability* (Section 1.6).

1.1 What Is Manufacturing Cost Policy Deployment?

Based on the assumption that the results are what really matters and that the consistent and constant customer satisfaction is the major concern of manufacturing companies, as with any other company, they are focused to continuously harmonize their products and production method to the market needs by continuously planning and controlling the transformation of the existing and future product processes in order to have an acceptable level of cost and flexibility and quality just as the rules of competition require. Further on, this process of harmonious and continuous transformation of resources into expected results requires a continuous adjustment of the current state by performing systematic (with small and continuous steps) and systemic (with large/radical steps and not on a continuous basis) improvements steadily directed toward the fulfillment of needs of the expected results.

1.1.1 Main Barriers to the Consistent Transformation of Manufacturing Companies

Achieving the expected results by using a consistent and harmonious transformation method of resources into outputs requires uniformity of the manufacturing vision across the company (with all its entities). Often this uniformity hits some barriers that obstruct the achievement of results (outputs) in line with what might be achieved with current facilities and the real needs of the market, or with what was planned. From the perspective of the need to continuously reduce unit costs and beyond, the main barriers to systematic and systemic improvement of the current transformation method are (1) The Lack of Real Managerial Commitment, (2) Resistance to Change, (3) The Lack of Total and Continuous Involvement of All Departments and Beyond, (4) Reactive Managerial Behavior, and (5) Incorrect and/or Incomplete Improvement Projects Implementation. In the following we will brief out these five main types of barriers.

1.1.1.1 The Lack of Real Managerial Commitment

The active and visible participation of the management team in establishing and supporting the strategic improvement directions and maintaining a continuous desire for change toward the betterment of the production method is still an important drawback for many management teams. The support for improvement often exists only at a declarative level. In fact, building a culture of improvement is considered more or less a fad, and consequently, companies do not have a vision, a mission, concrete goals, and robust plans for improvements in tandem with the company's business plan, even if they "make full use of the best methods, techniques, and tools for improvement." In this context, management teams sometimes do not have a level of strategic and comprehensive understanding of the improvement impact on the entire organization in the short, medium, and especially long term. The main consequences of the lack of this real managerial commitment is the lack of confidence in the ability to achieve consistent improvements, especially due to the failures of past improvement projects, lack of prioritization of improvements, inability to measure in advance the impact of the improvements in money, and, consequently, the lack of resources necessary for the improvement (especially the time needed for people to participate in improvement projects). In fact, managerial teams are focused almost exclusively on performing current activities and less on long-term health of the manufacturing system by making improvements. Managers are not fully aware that continuous development and implementation of improvement scenarios and the master plan also help them to achieve short-term performance targets.

1.1.1.2 Resistance to Change

The continuous adaptation of manufacturing companies to changing market needs and the continuous support for the transformation of the current production method require an awareness that improvements and, implicitly, ongoing change are part of the current and future life of the company. Not a few times, maintaining a new standard resulting from an improvement project lasts for a short period of time, or it is not even fairly communicated to those who should follow it in their daily tasks. Therefore, in order not to return slowly but surely to the old habits (possibly standards), the full commitment of those who implement the change of the new standards (rules), the allocation of resources needed for the improvements, effective communication, the full involvement of the managers in supporting changes, the development of training programs appropriate to current or future needs, robust knowledge transfer plans (with one-point lesson, OPL, techniques), the development of knowledge management (including skills and expertise), and the full understanding of cost vs. benefits from improvements reasoning are a must—if possible before improvements are made, etc. In fact, the easier acceptance of a change takes place if those who will have to accept the change (the new standard) participate in its design (the improvement projects) and if any change has a clear strategic motivation. The top management must continuously provide all those involved with answers to questions such as: *Why is the change implemented? What is the problem now? Why is the change needed now? Why is it us who must make the change and not others?* Therefore, to accept change as a normal and everyday state, there is a need for continuous reconciliation between the top-down approach (or desires, including the need for cost reduction) and the bottom-up approach (or possibilities, including real opportunities for *Manufacturing Cost Improvement,* MCI), using a *catchball process* based on managerial coordination according to predetermined targets. In the absence of this continuous reconciliation between desires and real possibilities, resistance to change occurs especially among middle managers, even if top managers and operators often desire to shift the current production method to the better.

1.1.1.3 The Lack of Total and Continuous Involvement of All Departments and Beyond

Making improvements is the task of everyone, just like performing current activities.

From the perspective of medium- and long-term profit targets (three to five years), the manufacturing companies have the following approach:

$$\text{Profit targets} - \text{External profit} = \text{Internal profit} \qquad (1.1)$$

From the perspective of the Manufacturing Cost Policy Deployment (MCPD) system, the *target profit* means the level of manufacturing profit required to satisfy the shareholders' need for profit and the need for sustainable development of the company. For example, the profit target level for a certain period can be $1 million. The *target profit* term can be extended to all dimensions of the company (supply chain, research and development [R&D], etc.). In this book, we refer to the manufacturing area.

External profit is understood as the expected medium- and long-term profit from sales of goods and services to the company's customers. For example, if the total revenue for a given period is $3.5 million and the total expenditure is $3 million, then the external profit is $500,000.

The *internal profit* means the necessary profit to be obtained through MCI. Continuing the earlier example, it is $500,000.

Therefore, the participation in making unnecessary resource use improvements is not just the "duty" of the production department and the direct support departments (especially the maintenance, quality, and engineering departments), where production takes place, but of all departments and all people involved (including from outside the company). Depending on the competitive pressure, the continuous improvement of the company's long- and medium-term business strategies is made for each competitor, for each customer, for each current or future product, for each supplier, for each process/work center, sometimes for each activity (such as setup activity), for each equipment/line, and for each department and employee. This way, each department (and then an employee or third person outside the company) must proactively contribute to supporting the fluency of the basic workflow by selecting, reviewing, and following up strategic improvement projects and by developing an organizational structure to facilitate in obtaining profit (external but mainly internal profit). Therefore, identifying means to reach predefined targets of different Key Performance Indicators (KPIs) to meet manufacturing companies' strategies is the task of each department and of every person involved in the company's activities. So, companies develop an annual action plan for systematic and systemic improvements, including MCI, fully involving all available workforce and all available facilities, as the fluency of the company's core flow is more important than the activities of any department or employee (including any manager).

1.1.1.4 Reactive Managerial Behavior

The real value of a manager is his/her ability to accurately plan the business to prevent potential problems from occurring. The continuous monitoring of tangible and especially intangible performance sometimes determines managers to make decisions at a time to quench "fires." This alert state can take much of a manager's time, and they "have no time" to make proactive and/or preventive decisions. The preponderance of such a managerial style to "solve the problems" at a time does not contribute to sustaining a culture of consistent improvements and, implicitly, to reducing and/or eliminating unneeded resource consumption. Often such a reactive managerial behavior has deficiencies in substantiating decisions based on actual data and facts and creates a state of confusion and continuous stress in companies. Therefore, developing a *desirable contextual management identity* (*management branding*) (Posteucă, 2011) within the managerial team is required to continuously improve managers' behaviors and to provide a creative atmosphere for smaller or larger improvements based on long-term productivity. The continued connection of the necessary improvements to the long-term strategies of the company requires increasing the level of stability and standardization of the production flow, and the use of a preventive and proactive decision-making style to meet the company's long-term goals.

1.1.1.5 Incorrect and/or Incomplete Improvement Projects Implementation

One of the main effects of the four barriers mentioned earlier is the level of effectiveness and efficiency of implementing improvement projects. The purpose of the improvements is to implement solutions to the root causes of problems, or solutions for the undesirable effects of a current state from processes visible at KPIs targets, at goals achievement and successful company strategy implementation. Often the members of the improvement projects teams have difficulties in understanding and precisely measuring the impact of improvements in current and future target performance of the manufacturing flow, the connections to company strategy, often lacking a coherent annual action plan. At the same time, many manufacturing companies are still focusing on systematic and systemic improvements almost exclusively in the production area (although some still have problems with a deep understanding of current processes) and, as a result, face multiple contamination of these improvements, as they do not have a *supply chain management* (SCM) approach. In fact, *improper implementation* of improvement projects is the result of poor understanding and/or misapplication of improvement methodologies, tools, and techniques, including *plan–do–check–act* (*PDCA*) methodology and/or *Define, Measure, Analyze, Improve, and Control* (*DMAIC*) methodology, which leads to insufficient skills to make improvements and to continuously support the necessary creativity. The *incomplete implementation* is the result of a lack of correlation of improvement projects with the company's strategy and especially with the company's strategy and profit plan, especially the internal profit plan. When it is not possible to scientifically predetermine the potential earnings in money of future systematic and/or systemic improvements, any cost of improvement implementations may become too high to be accepted. In conclusion, only the correct (consistent and effective) and full implementation (to achieve the company's strategies/internal profit plan or MCI, efficiently) of the improvement strategies can help sustain a robust and continuous transformation of the current manufacturing method in order to cope with a competitive global market.

1.1.2 Defining the MCPD System: Positioning and Description

Consequently, the consistent transformation of the manufacturing systems depending on market signals requires answers to the following questions: (1) *How can customers be satisfied and how can target profit be obtained on a continuous basis?* and (2) *What are the main problems of the current manufacturing method?* Customers' satisfaction addresses especially cost, quality, and delivery time levels (including the flexibility level). The order of importance of the three dimensions of a company's performance before the customer changes according to a number of current and/or future factors, such as current and future state of sales (growth or decrease;

e.g., due to lower sales, the cost dimension or the continuous reduction of costs becomes the most important factor), industry type (e.g., the quality will be the most important dimension in the electronics industry on a continuous basis), product type (e.g., the terms of delivery will be the most important for perishable foodstuffs), etc. Regarding the main problems of the current manufacturing method, knowledge of and ongoing approach to the *overcapacity* state (planned capacity utilization < practical production capacity) or the *undercapacity* state (planned capacity utilization > practical production capacity) or the *undercapacity with a hidden overcapacity* state are necessary. Manufacturing companies need to constantly and in depth become aware of their current capacity and future capacity needs so they can harmoniously plan the necessary transformation.

Normally, to achieve the target profit by increasing the external profit, it is necessary to increase the price, the sales volumes, or both. Often the increase of the sales price in the market is quite difficult to achieve. Then, increasing sales volumes based on increased production volumes require increased capacity by (1) increasing current capacities, which often cause overloading of equipment and/or people, which promotes the occurrence of waste and losses and implicitly increased associated costs not found in customer value, and (2) increasing capacity through investment, which requires a certain amount of time to become available even more a certain investment cost to be supported by the price level. Thus, the increase in production volumes will result in an increase of variable costs, both direct (raw materials, materials, and direct labor) and indirect (mainly associated to losses and waste), but also of fixed costs related to potential new investments for capacity increase. In this context, to achieve the target profit level, meeting internal profit with MCI is sought, based on the reduction of cost of losses and waste (CLW) scientifically and continuously measured.

Moreover, current capacities are often diminished due to the occurrence of losses and waste (inventories) throughout the entire manufacturing flow. Transforming losses and waste (inventories) into costs in order to establish opportunities for MCI is a way of continuously satisfying customers, both in terms of costs and implicitly price, and from the perspective of targeting the capacity adjustments to ensure the volumes of current and/or future products and/or services by implementing systematic and systemic improvements that are continually coordinated by signals coming from competitors, customers, suppliers, and especially shareholders (the need for long-term profit).

In this context, the concept of *Manufacturing Cost Policy Deployment* (MCPD) is defined as the process of translating the strategic objective of reducing production costs in the long run toward the improvement of annual systematic activities and toward annual systemic improvement actions by setting targets and means to improve process costs of families of products, in order to

1. Fulfill the annual improvement budget (both existing products and new products), by continuous investigation of the relationships between costs, processes, and losses and waste.
2. Achieve performance of annual cash improvement budget.
3. Direct and plan the systematic and systemic improvements through continuous reconciliation between the need to reduce costs and opportunities for MCI.
4. Measure and analyze performance for MCI.
5. Achieve cost targets at shop floor level (Posteucă, 2015; Posteucă and Sakamoto, 2017, pp. 81–82).

From the perspective of MCPD, the concept of policy refers to the setting of targets and means. Akao says: "target and means combined can be called a policy" (Akao, 1991, p. 5). The other concepts in the MCPD definition, such as the *annual improvement budget*, will be described in the following chapters.

Figure 1.1 shows the positioning of the MCPD system within the manufacturing companies in order to support their transformation coordinated by the need to continuously reduce the costs imposed by customers, competitors, and shareholders (the approach from the outside to the inside, or the *demand for the reduction of unit manufacturing costs*) and the need for MCI (the approach from the inside to the outside—*the offer of reducing the unit manufacturing costs*) by systematically and systemically improving the level of productivity and quality of the manufacturing flow for current and/or future products. Reducing and/or eliminating the five barriers described earlier are the basic purposes of the three phases of the MCPD system.

Therefore, in synthesis, the three phases of the MCPD system are as follows:

■ Phase 1. Manufacturing cost policy analysis
  Purpose: Understanding the connections between costs, losses, and waste and cost reduction strategy for each product family to ensure the awareness of the need for continuous transformation of manufacturing companies in order to increase the *real managerial commitment* and to mitigate and/or eliminate the *resistance to change* of all the people in the company

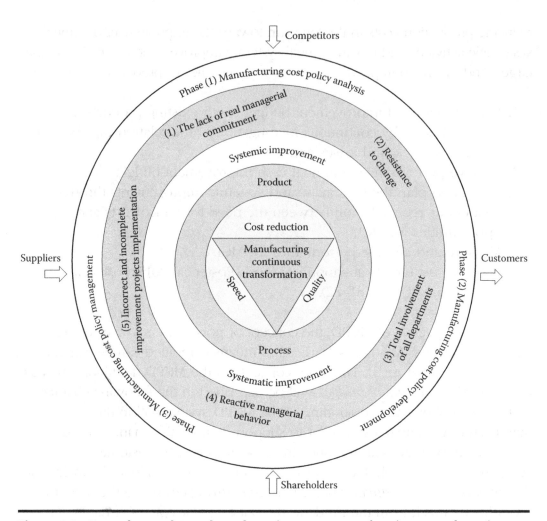

Figure 1.1   From the need to reduce the unit costs to transforming manufacturing companies through the three phases of the MCPD system.

Costs output: Annual targets and means for manufacturing costs behind losses and waste (MCI targets and means/MCI policy deployment)

Approach: Defining the current and future states of productivity (productivity vision, productivity mission, productivity strategies, and *productivity core business goals—PCBG*)

■ Phase 2. Manufacturing cost policy development

Purpose: *Annual manufacturing improvement budgets* to support the continuous targeting of targets and means in order to meet the long-term profit plan of both the external profit (from the sale of goods and services) and especially the internal profit (from systematic and systemic improvements of the manufacturing processes) through *total involvement of all departments* to achieve the continuous company transformation and by reducing *reactive managerial behavior* to operational challenges based on the annual action plan for MCI

Costs output: Annual action plan for MCI

Approach: Cascade MCI activity plans

■ Phase 3. Manufacturing cost policy management

Purpose: Engage the workforce to execute MCI strategies to reduce or eliminate *incorrect and incomplete improvement projects implementation* by continuously targeting these improvement projects through the need for unit cost reduction imposed by the market

Costs output: Cost management performance improvement and daily cost management (DCM)

Approach: Structured meeting for the MCI

1.1.3 MCPD System: Principles and Basic Features

Therefore, the key objective of the MCPD system is to achieve the harmonious transformation of the manufacturing companies by coordinating this transformation in line with the developments of the need for cost reduction imposed by the market (shareholders, customers, competitors, and suppliers). Seven basic principles have been developed to make it easier to understand the core purpose of the MCPD system, namely, to achieve the continuous transformation of manufacturing companies to ensure their synchronization with market needs by aligning all systematic and systemic improvements coordinated by the need of profit and continuous cost reduction (Table 1.1) (Posteucă and Sakamoto, 2017, pp. 82–86).

Table 1.1   The Basic Principles of the MCPD System

Principle no. 1	Target profit from MCI does not change.
Principle no. 2	MCI targets for each PFC.
Principle no. 3	The continuous quantifying of losses and waste in costs for each PFC.
Principle no. 4	The continuous reconciliation of annual MCI targets for each PFC.
Principle no. 5	Improvement budgets for each PFC.
Principle no. 6	Coordination improvements through MCI targets for each PFC.
Principle no. 7	Waste (stocks) elasticity on losses.

The key element of these principles is that the MCPD system must reflect in a predetermined way the economic need for profit by continuously reducing unit costs and not just the need for speed and quality of processes or, in other words, the long-term relevance of profit against productivity and quality. Essentially, this way of seeing improvements through the need for continuous reduction of unnecessary costs is based on the premise that the need for profit is perennial for the survival and development of the manufacturing companies. The essence of reducing and/or eliminating losses and/or waste (stocks) is to generate profit, both external profit through capacity increases and, above all, internal profit by reducing unnecessary costs. Therefore, in terms of the MCPD system, productivity is addressed both in terms of outputs (focusing mainly on increasing capacity, especially if sales are increasing) and inputs (focusing mainly on cost savings, especially if sales are declining).

Table 1.2　The Basic Features of the MCPD System

Goal	Continuous fulfillment of the long-term company profit plan, especially domestic profit, and the successive reduction of unit manufacturing cost through MCI by continuously improving the productivity of existing and future products
The guiding principle	The relevance of long-term profitability to productivity and quality
Normal condition	Scientific measurement and continuously improving the cost behind losses and waste from all manufacturing processes for each PFC
Focus of improvement	MCI by improving losses (time-related losses and physical losses) and waste (stocks) for each PFC
Working method	Establishment of manufacturing cost policy (MCI targets and means) by top managers through the continuous involvement of all the people in the company and beyond
The main managerial interest	Continuous support of the multiyear profit plan with the continuous improvement of productivity
Desirable output	Outside the company: profit and price competitiveness Inside the company: continuous and harmonious transformation of the manufacturing flow

In the MCPD system logic, the costs behind losses and waste (stocks) are the ones that decide the assessment of the timeliness of systematic and/or systemic planning. For this, we continuously quantify the costs of losses and waste (stocks), especially for every bottleneck along the entire production flow for each product family, so that managers become aware of the real impact on costs and implicitly on profit of losses and waste and not just their impact on lost times (such as line unbalancing), non-quality (rework and scrap), and inventories (such as work in progress, WIP).

In this context, the basic features of the MCPD system are shown in Table 1.2.

In conclusion, from the perspective of the MCPD system, the continuous and harmonious transformation of the manufacturing companies should be driven by the need to meet the long-term profit plan through systematic and systemic productivity improvements, by ensuring an organizational culture based on flexibility and openness to changes imposed by the market.

1.2 From Continuous Cost Reduction to Manufacturing Cost Improvement

Continuing the logic of Formula 1.1 at the level of one year, one can say that

$$\text{Annual target profit} - \text{Annual external profit} = \text{Annual MCI goal} \quad (1.2)$$

The annual MCI goal means the level of all cost improvements in the manufacturing area that must be achieved at the level of the group of companies to meet the annual profit gap between the target profit and the external profit (from sales of goods and services to customers).

Therefore, the continuous manufacturing cost reduction is and will remain the major goal of managers to achieve long-term profit, by internal profit (see Formula 1.1), especially as multiple and accentuated variations of sales volumes occur, resulting implicitly in a variation of the current manufacturing capacities. Variations in manufacturing capacities have an impact on the manufacturing cost, both in the case of *undercapacity* by overburdening current resources (especially equipment and people), which results in variable cost increases, especially in manufacturing overhead, and also in the *overcapacity* state by the nonuse of resources that are highlighted in the high costs of products, especially fixed costs.

Regarding the *undercapacity with hidden overcapacity* state, a common condition in many manufacturing companies, all cost structures tend to be oversized as the current manufacturing flow capacity is diminished by multiple constraints:

■ *Losses* (which largely refer to the transformation times): seven equipment losses (*failure losses, setup/adjustment losses, cutting-blade losses, start-up losses, minor stoppage/idling losses, speed loss, defects and rework losses*), five human work losses (*management losses, motion losses, line organization losses, losses resulting from failure to automate, measuring and adjustment losses*), and three production resource losses (*yield losses; energy losses; die, jig, and tool losses*) (Nakajima, 1988; Shirose, 1999, pp. 40–61)
■ *Waste*: mean stocks, which refers to material elements (Ohno, 1988, pp. 19–20; Posteucă and Sakamoto, 2017, pp. 28–30)

Going forward, the annual MCI goal group-wide is allocated to each company within the group. The annual MCI goal group-wide is

Annual MCI Goal group-wide = Annual MCI Goal for Plant 1
+ Annual MCI Goal for Plant 2 + Annual MCI Goal for Plant 3 + ⋯
+ Annual MCI Goal for Plant "n" $\qquad$ (1.3)

Further on, the annual plant MCI goal is allocated to each product family cost (PFC) (PFC means product families that run around the same processes, have roughly the same MCI opportunities, and have the same MCI target—usually the target of the products with the highest volumes under the PFC cost and run around most, if not all, processes). So, the annual plant MCI goal is

Annual MCI Goal for Plant 1 = Annual MCI Goal for PFC 1
+ Annual MCI Goal for PFC 2 + Annual MCI Goal for PFC 3 + ⋯
+ Annual MCI Goal for PFC "n" $\qquad$ (1.4)

Furthermore, the annual MCI goal for each PFC is allocated over each process. The annual MCI goal for each PFC is

Annual MCI Goal for PFC 1 = MCI Targets Process 1
+ MCI Targets Process 2 + MCI Targets Process 3 + ⋯
+ MCI Targets Process "n" $\qquad$ (1.5)

The logic of thinking about the MCPD system in addressing the costs behind losses and waste and, implicitly, establishing *MCI targets and means* is that of *crosscutting causality* along the entire production flow and beyond. For example, a reduction in the current capacity of processes by increasing losses, for example, the decrease in the planned percentage of Overall Equipment Effectiveness (OEE) for a strategic equipment deemed as bottleneck, has the effect of increasing waste, especially WIP for the area, but also other losses in the process and along the manufacturing flow, such as human work losses, but implicitly with effects on manufacturing costs for the entire manufacturing flow and beyond.

The setting of *annual MCI targets* depending on the current and immediate situation of the manufacturing flow is based on the total level of *costs of losses and waste* (CLW) (*chronic cost problems*) of all processes/work centers (transforming losses and waste into costs) and especially based on the total level of *critical cost of losses and waste* (CCLW) (*acute cost problems*) from critical processes/activities (by summing up the impact of the *root cause of CLW* over the entire manufacturing flow and beyond). The setting of annual MCI means is based on the level of losses and waste for each process/work center and especially on the basis of each process/work center that generates losses and waste along the entire manufacturing flow and beyond. Therefore, meeting the MCI targets (addressing the effect) depends on the consistent approach of cost causes (CLW and especially CCLW), and further on by consistently addressing process/work center causes, the current level of losses and waste, through continuous improvement activities (kaizen) and systemic improvement actions (kaikaku).

Therefore, following the logic of Formula 1.5, we have

$$\text{MCI targets process 1} = \text{CLW targets from process 1}$$
$$+ \text{CCLW targets from process 1} \qquad (1.6)$$

The annual MCI goal for PFC1 is met by meeting the CLW targets of each process related to PFC and by meeting CCLW targets both at the process level and especially throughout the manufacturing flow and beyond (collects the costs of losses and waste from several processes that are affected by an acute manifestation of losses or waste).

Moving forward, building the long-term profit plan for companies (manufacturing profit targets) and work volumes often start from the three basic customer needs: cost/price, quality, and delivery times. The cost, as an entity both internal and external to the company (through price and profit), is the one that influences consistent profits over time. In this regard, Ohno (1988, p. 8) states that "Efficiency, in modern industry and business in general, means cost reduction. (…) At Toyota, as in all manufacturing industries, profit can be obtained only by reduction costs. When we apply the cost principles selling price = profit + actual costs, we make the customer responsible for every cost." However, Imai considers that cost reduction is not the basic purpose of gaining profit by increasing efficiency, but rather *cost improvement*: "the word cost does not mean cost cutting, but cost management. Cost management oversees the processes of developing, production, and selling products or services of good quality while striving to lower costs or hold them to target levels" (Imai, 1997, p. 44).

By extending the *cost principle of Toyota* (*sale price − profit = costs*) (Shingo, 1989, p. 75; Hirano, 2009, pp. 38–39), one may deem that

$$\text{Price} - \big(\text{External profit} + \text{Internal profit}\big)$$
$$= \text{Value-added costs} + \text{Non-value-added cost} \qquad (1.7)$$

The external profit is the profit obtained from the sale of goods and services to customers. The internal profit is understood as the profit obtained through the continuous improvement of manufacturing costs, especially the non-value-added cost. The non-value-added cost represents the quantification of losses and waste in costs at the process level of each PFC.

For example, as the sales price or the external profit (sales volume) cannot increase, if the current sales price is $100, the current external profit is $7, the actual value-added costs is $76 and the actual identified non-value-added cost is $17 and a target profit of $15 per unit is required, then the $8 non-value-added cost reduction is needed to achieve a $15 unit target profit.

In this context, obtaining profit by cost reduction as Ohno suggests is mainly transformed into continuous non-value-added cost improvement by continuously improving losses and waste underlying these costs for each manufacturing process of each PFC or product with robust cost management invoked by Imai. In this respect, the continuous reconciliation between manufacturing target profit and target costs, especially through the achievement of an internal target profit with the possibility of continuous non-value-added cost improvement, represents the *polar star* of all productivity improvements in the manufacturing flow from the perspective of the MCPD system. Through the drive effect, this coordination of productivity improvements will also have an impact on releasing current capacity that will have an impact on the increase in production volumes and sales, which will also generate external profit. So, the major challenge is to identify and reduce the non-value-added cost by identifying and reducing associated CLW and CCLW (through MCI).

The continuous non-value-added cost improvement covers all costs. From the manufacturing cost perspective, these costs aim at *transformation cost* (direct labor costs, indirect labor costs, manufacturing overhead, and depreciation costs) and *material costs* (direct material costs and indirect material costs/auxiliary materials cost). Furthermore, MCI aims especially at *transformation cost improvement* (in particular, direct labor, production overhead—maintenance costs, utility cost, and less depreciation costs), but also the consumption of raw materials and auxiliary materials. Reducing the cost of materials as a result of price reductions through successive negotiations with the suppliers does not fall within the scope of manufacturing cost improvement. It falls within the scope of *supply chain management* (*purchasing*). At the same time, the reduction of costs of raw materials and auxiliary materials through the design of new products does not fall under the scope of MCI. It lies within the scope of R&D. From the perspective of new products, MCI includes non-value-added cost related to losses and waste associated to future processes within the manufacturing flow that are similar to those of existing products in production and for which activities and/or actions to improve productivity are already in progress.

In conclusion, the main challenge of obtaining the annual manufacturing target profit through meeting the annual MCI goal and unit manufacturing target profit is to locate, calculate, and improve CLW and CCLW by productivity-enhancing methods, techniques, and tools by involving everyone in and beyond the company based on a coherent cost strategy.

1.3 Cost Strategy into Action: Top-Down and Bottom-Up Approaches

To fulfill their vision and productivity mission (Posteucă and Sakamoto, 2017, p. 21), starting from the need to continuously meet the long-term profit plan, companies are developing PCBG and cost strategies accordingly. These strategies are detailed for each product family and start from the current and future status of each product of the product family (profit level, price level, cost level and structure, product life cycle, sales volumes, level of CLW and CCLW along processes, etc.).

From the perspective of the MCPD system, there are two pressures that are constantly reconciled:

■ *Top-down approach*: The need for meeting the annual MCI goal imposed by the market (stakeholders, customers, competitors, suppliers), supported by senior managers in particular in order to obtain a competitive price and target profit

■ *Bottom-up approach*: The need for meeting MCI targets, supported by middle managers in particular, through continuous improvement of CLW and CCLW at the level of processes related to each PFC

Therefore, in the MCPD system paradigm, *cost strategy into action* is a systematic planning process that aims to align the continuous cost reduction strategy with the daily activities of systematic and systemic improvements of MCI through CLW and CCLW improvements across all processes of a PFC. To successfully implement the cost strategy, all people in and beyond the company need to know their role in continuously transforming the company and meeting the PCBG.

The cost strategy development and furthermore the productivity strategy development target three interconnected thinking levels:

1. *The basic cost strategy* is determined by the status of sales volumes, their decrease or increase, respectively, and further on the overcapacity or undercapacity state. Therefore, in the context of a decline in sales volumes and an increase in overcapacity, it is necessary to increase the internal profit through MCI, in particular, by reducing costs such as labor, raw materials and auxiliary materials, maintenance, and utilities. In the context of an increase in sales volumes and increasing undercapacity, it is necessary to achieve internal profit by unlocking current capacities and associated manufacturing cost, in particular, by reducing the cost behind lost time of equipment and people, behind the differences between the standard cycle time and the actual cycle time of the equipment, and behind the non-quality time. The basic cost strategy is designed at the company level for a period of three to five years depending on the sales trend and determines the necessary productivity strategy and the master plan of improvements.

2. *The interdepartmental cost strategies* are the strategies that are developed within each department, in an interdepartmental approach, to coordinate all the activities and actions necessary to fulfill the MCI

targets. For example, in the downside of sales volumes, having as basic strategy the cost reduction for auxiliary materials, to reduce the consumption of soldering tape (auxiliary material cost) in a process in the production area, it is necessary to involve several departments such as production, maintenance, quality, internal logistics, and HR.

3. *The PFC strategies* are the most visible cost strategies because they are at the level of each relevant product and rely heavily on the life cycle of products. From the perspective of the MCPD system, they aim at identifying all the cost reduction opportunities using MCI. This strategy is developed by continuously capturing market signals (customers, suppliers, competitors, and shareholders) and transmitting signals about the level of MCI targets required at product level. Furthermore, based on the required MCI targets, MCI means are developed by identifying the most feasible opportunities to reduce CLW and especially CCLW in the processes of each product family, by involving all the people in the company and beyond.

MCI targets and means (MCI policy deployment) are established with the help of the three types of strategies, and they serve as genuine transformation triggers for MCI at the level of the entire manufacturing flow.

1.4 MCI Policy Deployment: Targets and Means

Unlike cost strategy, which is a multiyear plan to reduce costs through strategic resource planning and mobilization, MCI policy deployment is the presentation of both MCI directions and MCI control management to reach MCI directions for shorter periods of time (usually annually, with a six-month revaluation). However, cost strategy often develops and adjusts concurrently with the establishment of MCI policy deployment. MCI policy deployment coordinates the company policy deployment to meet the long-term profit plan with the help of productivity increase, through systematic and systemic cost improvements in each of the two possible business scenarios:

1. *In the case of sales growth*, profit is achieved by maximizing the manufacturing capacity (stressing the obtaining of external profit predominantly by maximizing the effectiveness) and by limiting unnecessary resource consumption (maintaining inputs constant; not effectively used input or *losses*).

2. *In the case of decrease in sales*, profit is obtained by minimizing inputs, especially raw materials, auxiliary materials, and direct labor (stressing the obtaining of internal profit predominantly by maximizing efficiency), and limiting unnecessary resource consumption (maintaining outputs constant; limiting the excess amount of input or *waste*).

Therefore, in the sense of MCI and implicitly of MCPD, profit making is achieved by obtaining both external and internal profit. The share of the two types of profit is determined by the current and future state of sales. The main goal of MCI and MCPD is to continuously ensure an acceptable level of target profit by continuously exploiting all opportunities to obtain internal profit by reducing costs behind losses and waste from processes. Over time, the share of internal profit in the company's target profit tends to fall due to the reduction and/or elimination of MCI opportunities, especially through systematic improvement activities (kaizen).

MCI policy deployment is in fact the setting of MCI targets (or expected outcomes for CLW-related KPIs and, in particular, CCLWs or operational directions required to meet the cost strategies outlined earlier) and MCI means (or targeting/concrete means to achieve MCI targets/directions for implementing past cost strategies through a *performance cost management control*, specifically by addressing losses and waste and critical losses and waste). Setting targets (desired results or effects) and means (causes) for MCI can be considered as two facets of the same coin. Establishing MCI targets is done at the same time as establishing the MCI means. Sometimes, setting targets and means for MCI can be done separately, but they must be linked to one another—the sooner the better. If MCI targets were set without considering at all the real MCI means, unrealistic MCI targets could be set (too easy or too difficult/impossible to reach, both leading to the demoralization of people and weakening, sooner or later, of the desire to make consistent and continuous transformations at process level). In the same logic, establishing MCI means without considering MCI targets is the way to carry out activities and actions of improvement without effectiveness and efficiency.

The intensity of MCI targets or the difficulty of reaching MCI targets determines the level of resources required to implement MCI means. A nonperforming way to establish MCI targets is to leave the MCI's annual percentage of improvement to the discretion of managers. They will find countless apparently objective justifications for a modest percentage of the annual MCI targets. Another, equally unfortunate, way is to establish MCI targets perceived as hard to achieve on the background of an accurate lack of localization of processes and activities with the greatest and the most at hand potential to achieve the MCI targets for a PFC or another. In this context, without a clear targeting of how to meet MCI targets, often not enough resources are allocated to MCI means (improvements) and the transformation of manufacturing companies takes place in waves (when resources are available, improvements are made, when not they drop on the second place, and operational activities are paramount) and the results are inconsistent over time. Therefore, the best way to set MCI targets is to clearly show the processes that can support MCI targets and to choose the most efficient and effective MCI means to achieve the unit manufacturing cost reduction during the time limit imposed by the competitive game. This way, the MCPD system provides the reaction speed needed to achieve a consistent manufacturing cost reduction at the right time.

In order to direct MCI targets and means at a PFC process level and to reach the CLW and CCLW approaches for every process of each PFC, it is necessary to go through the next ten preceding steps (connections or stratifications) (Posteucă and Sakamoto, 2017, pp. 140–154):

1. Vision, mission, and value vs. total manufacturing costs (history and future constraints)
2. Total manufacturing costs vs. overall management indicators (OMIs)
3. OMIs vs. KPIs for each PFC
4. KPIs for each PFC vs. KPIs of losses and waste for each PFC (Posteucă and Sakamoto, 2017, pp. 119–122)
5. KPIs of losses and waste for each PFC vs. critical KPIs of losses and waste for each PFC
6. KPIs of losses and waste for each PFC vs. CLW for each PFC
7. Critical KPIs of losses and waste for each PFC vs. CCLW for each PFC
8. CLW for each PFC vs. CCLW for each PFC
9. CCLW and CLW for each PFC vs. MCI targets for each PFC
10. MCI targets for each PFC vs. MCI means for each PFC (for step 5)

Therefore, the entire construction of the OMIs for productivity, quality, delivery, safety, and people's morale must serve to ensure a long-term acceptable profit (long-term profit plan/target profit) and rely in particular on achieving an acceptable level of unit costs (costs that would ensure a competitive price level and, moreover, a competitive profit level). Behind the unit costs lies mainly the level of productivity and quality in the processes. Furthermore, the necessary cost level directs the required level of achievement of all KPIs by quantifying losses and waste in costs (Posteucă and Sakamoto, 2017, pp. 116–140).

Figure 1.2 shows the four drives through which annual MCI policy deployment is developed (targets and means connections for each PFC):

1. *Market driven for MCI*: The board of directors and senior managers analyze the historical trend and the main potential future constraints for *sales turnover* (production number, average price evolution, stock level of finished products, units sold daily, etc.) and *market share for*

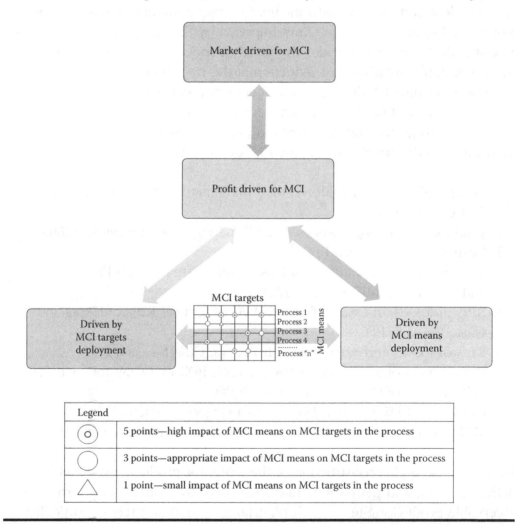

Figure 1.2   MCI policy deployment: targets and means connections for each product family.

target area. From the perspective of MCI, *average price evolution* is the main area of analysis and interest to determine the level of potential target profit and manufacturing cost.

2. *Profit driven for MCI*: The board of directors and senior managers analyze the historical trend and the main potential future constraints for the *annual manufacturing target profit, annual manufacturing unit target profit, annual manufacturing internal target profit* (percentage of manufacturing internal profit from kaizen in annual target profit; number of kaizen projects for MCI in all departments—before/after kaizen, office kaizens, number of people in kaizen, number of kaikaku projects for MCI), *annual manufacturing target profit from new products* (especially annual MCI goal from kaizen for new products), *scrap ratio, rework, production capacity*, etc. From the MCI perspective, *annual unit manufacturing target profit, annual manufacturing internal target profit*, and *annual manufacturing target profit from new products* are the main areas of analysis and interest in establishing MCI targets. In order to establish the realistic profit level for existing and future products throughout their entire life cycle, profit driven for MCI reconciles with MCI targets through the catchball process to determine the real possibilities of annual reduction of manufacturing cost for each PFC.

3. *Driven by MCI targets deployment*: Senior managers, middle managers, and MCPD implementation team analyze the historical trend and the main potential future constraints for the following:

   ■ *Total manufacturing cost (initial budget)*
   ■ *Percentage of annual unit manufacturing cost reduction*
   ■ *Manufacturing unit cost (standard)* (transformation costs, material costs, depreciation cost; determining the cost of a minute of activity for each product type in each process/work center)
   ■ *CLW* for each process/work center for each PFC or product
   ■ *Total costs of losses and waste* for all products (of the standard cost; the total offer for the MCI for equipment, human work, materials/energy, and waste/stocks)
   ■ *Percent CLW from unit manufacturing cost* (average unit for equipment, human work, materials/energy, and waste/stocks)
   ■ *CCLW* (by summing up the impact of the *root cause of CLW* throughout the manufacturing flow and beyond)
   ■ *Percent CCLW from unit manufacturing cost*

Based on these analyses, the *annual MCI goal for the entire company and for each PFC* and *MCI targets for each process of each product and/or PFC* are set depending on the CLW and CCLW percentages of unit manufacturing cost and taking into account the required level of cost reduction established by *profit driven for MCI target profit.*

At the same time, *annual MCI targets deployment for CLW and CCLW chosen to be improved* are achieved (*especially for CCLWs that have a greater impact on MCI*) for each of the following:

■ *Work center/area, processes, and activity*
■ *CLW and CCLW* (equipment, human work, materials/energy, and waste/stocks)
■ *Important customer* (to meet the demands of successive price cuts)
■ *Representative product* (to meet the successive need to reduce the unit manufacturing cost)
■ PFC
■ *Group leader,* usually department manager depending on the preponderant CLW types chosen to be improved

From the MCI perspective, *annual goal for manufacturing cost reduction per unit/PFC and annual MCI targets on each area and process* are the main areas of analysis and interest in establishing MCI means for each process. To determine the realistic level of the annual MCI target at every process or important process level for existing and future products, MCI targets are reconciled with MCI means at process level through the catchball process to determine the real possibilities of annual reduction of manufacturing cost for each PFC.

4. *Driven by MCI means deployment:* Senior managers, middle managers, and MCPD implementation team analyze the historical trend and the main potential future constraints for the following:

■ *Manufacturing lead time* (number of workstations, total cycle time, man*hours by product, OEE, WIP inventory from setup, WIP inventory from moves, overall line effectiveness for assembly line, times for material handling, assembly line speed, etc.)
■ *Production delivery performance* (On-Time In-Full—OTIF; materials stock days, finished products stock days)
■ *Accidents*

From the perspective of MCI, *manufacturing lead time* represents the main area of analysis and interest in establishing on a continuous basis the difference between the manufacturing cost of each product family from the ideal state of manufacturing flow (the state when *one-piece flow* [OPF] and *continuous flow* principles are implemented completely along the entire manufacturing flow) and the current state of the manufacturing cost, with the current manufacturing lead time and the current level of timing to market needs (takt time for each product family). The comparative analysis of the ideal state with the current state identifies *weaknesses* for the following:

■ *Manufacturing key points* (MKP) for each product family: For example, the total six-month breakdown time (determined by event frequency and average duration of an event) from a process/work center/area and the impact of this breakdown time in all losses and waste influenced upstream and downstream of the entire manufacturing flow. It will seek to identify the process/work center/area/critical activities that cause other losses and waste along the entire manufacturing flow and beyond. The identification of MKP or of these critical points is based on the continuous measurement of the productivity level, namely, the level of losses and waste (Posteucă and Sakamoto, 2017, pp. 119–122).

■ *Manufacturing cost key points* (MCKP) or the cost behind every weakness (CLW and CCLW): Continuing the earlier example, with a six-month breakdown, one may determine the total cost of the breakdown for the directly affected process/work center/cost and the resulting cost by summing up all CLW throughout the entire manufacturing flow.

Knowing and fully understanding the connections between MKP and MCKP is the major challenge in determining MCI targets and means. Addressing this challenge takes place through realistic reconciliation between MCI means and MCI targets at every process level for products and/or PFC through the catchball process. The most feasible MCI means at that time are chosen through this reconciliation, which will meet the need to reduce the unit manufacturing cost of products in a product family by reaching MCI targets at the level of each manufacturing process.

In this context, MCI targets for the entire process of a PFC (e.g., number 1) becomes

$$\text{MCI targets for all process of PFC } 1 = \text{MCI means for process } 1$$
$$+ \text{MCI means for process } 2 + \text{MCI means for process } 3 + \cdots$$
$$+ \text{MCI means for process "n"} \tag{1.8}$$

Going forward, it is necessary to establish mean targets to meet MCI means for a process. For example, if it is considered necessary to meet the CLW target in a process and the CLW is formed from costs of setup losses, then it is necessary to have a target for costs of setup losses. To accomplish this goal, you need to have a setup time target (e.g., a target setup time reduction target from 25 to 9 minutes). In this context, MCI targets for all processes of a PFC (e.g., number 1) becomes

$$\text{MCI targets for all processes of PFC } 1$$
$$= \text{MCI means target deployment for process } 1$$
$$+ \text{MCI means target deployment for process } 2$$
$$+ \text{MCI means target deployment for process } 3 + \cdots$$
$$+ \text{MCI means target deployment for process "n"} \tag{1.9}$$

It should be stressed out that a high time (e.g., a 25-minute setup time) will not always automatically incur the most costs. The same 25 minutes for a breakdown, for the same equipment, with approximately the same impact on manufacturing lead time, may have double or even triple costs, if the breakdown event involved a higher cost associated with scrap and/or rework, than the costs generated by the set-up event (without scrap and/or rework). Therefore, a cost-oriented targeting is more accurate than an exclusive time and/or quality-oriented targeting, since sooner or later, customers in many industries and final customers are putting increasing pressures on cost reductions. At the same time, the continuous quantification of losses and waste in costs and continuous targeting of systematic improvement projects (meaning kaizen) and of systemic improvement projects (meaning kaikaku) ensures an increase in the ease of feasibility analysis of potential improvements and, implicitly, an easier allocation of resources for improvements.

In this context, MCI means targets deployment for a process (e.g., process 1) becomes

$$MCI\ Means\ Targets\ Deployment\ for\ Process\ 1...\ "n"$$
$$= MCI\ Means\ Deployment\ (level\ 1,\ PST)$$
$$+ MCI\ Means\ Deployment\ (level\ 2,\ kaizen)$$
$$+ MCI\ Means\ Deployment\ (level\ 3,\ kaikaku) \qquad (1.10)$$

MCI means deployment (level 1, PST) is the day-to-day planning of problem-solving projects related to MCI targets at process level (top five issues addressed at the shop floor level by operators, especially with the help of various problem-solving techniques—usually the A3 technique). MCI means deployment (level 2, *kaizen*) is the biannual planning of strategic systematic improvement projects (July of the current year and January of the following year; rectifying the annual action plan). MCI means deployment (level 3, *kaikaku*) is the annual and multiannual planning of systemic/radical improvement projects (new equipment, new technologies, new processes, etc.).

Therefore, the continuous and harmonious transformation of manufacturing flow is achieved through the two types of reconciliations: (1) between profit driven for MCI or MCI goal and MCI targets and (2) between MCI targets and MCI means through catchball process at all organizational levels. In fact, the need for cost reduction is being negotiated with the real cost reduction possibilities of the processes and activities of each product family, possibilities that are scientifically, systematically, and systemically identified.

Setting MCI targets and means (policy deployment) represents the key concept of the MCPD system.

1.5 Example of MCI Policy Deployment

Therefore, MCI policy deployment (targets and means) at process or PFC level is the extension of (1) the long-term profit plan, especially the internal profit plan obtained by improving productivity in the manufacturing area, (2) the medium-term manufacturing cost strategy, and (3) the need to reduce manufacturing unit cost in the short term.

In the following we will show an example of establishing MCI targets and means at AA-Plant. AA-Plant is a multinational firm operating in the automotive industry. The type of production at AA-Plant is *manufacturing and assembling.* AA-Plant is one of the top five companies within the group, with approximately 1250 employees. AA-Plant has three product families (PF, PF2, and PF3).

We will present the setting of MCI targets and means for PF3 (pilot project). For this pilot project, a team of seven people from AA-Plant (production manager, maintenance manager, quality manager, engineering manager, cost and budget manager, production planning specialist, and continuous improvement manager) was established, plus a consultant in the MCPD system. Champion was the plant manager at AA-Plant. The pilot project for the setting of MCI targets and means was part of the pilot project to implement the MCPD system for PF3, which took place over a period of 15 months. The pilot project for developing and setting MCI targets and means lasted about eight weeks.

The agreed objectives of the pilot project at PF3 by setting MCI targets and means were as follows:

1. Identifying the *annual MCI goal of $500,000 for the next 12 months* by implementing MCI means (meeting MCI targets by implementing the solutions identified through systematic improvement projects, kaizen, and systemic improvement projects, kaikaku)
2. *Reducing unit costs by 6% after eight months* (by meeting the $500,000 required to be obtained through kaizen and eventually kaikaku projects, and thus by reducing unnecessary expenses underlying losses and waste, CLW and especially CCLW, distributed at the level of planned and sold production volumes) based on price pressure imposed by customers
3. *Identifying process-level weaknesses* (MKP and MCKP) in order to establish the most effective and efficient MCI means
4. *Locating every opportunity for MCI* at process and activity level, and developing a mechanism for continually collecting data and information on current and future opportunities for MCI (CLW and CCLW)
5. *Choosing opportunities for MCI* to balance the demand for manufacturing cost reduction from customers through price with the opportunities for MCI at PFC level—possibly at product level, through the two types of reconciliations: (1) between profit driven for MCI and MCI targets and (2) between MCI targets and MCI means through catchball process at all organizational levels

6. *Setting acceptable MCI targets* from the perspective of those who will have to meet them, through catchball process at all organizational levels
7. Setting up the deployment for improvement projects (MCI means) and the implementation of best-in-time solutions in order to timely meet the need for a 6% reduction in manufacturing cost over eight months (*action plan*)

Following the MCPD system training, the project team was already aware that not only the most feasible MCI projects (as absolute values) were tracked but also the most feasible within the required cost reduction timeline. The eventual difference in unfulfilled feasibility of the improvement will be recovered from the profit gained from the cost reduction and the potential increase in customer demand on the background of a lower price. Any cost reduction delay may make MCI useless and, implicitly, unfeasible (resources consumed for improvements are not covered by a customer-centered effect at the price level).

To establish and develop MCI targets and means for PF3, the team has performed the following preliminary activities:

■ *Analyses of historical evolutions (representative years; usually the last three years) and forecasts of the need for cost reduction imposed by the market and in line with the productivity outlook*: The current and future average price level and structure for each PF1 product (price structure: % of unit costs, % of transformation costs, % of material costs), current profit level vs. the one necessary for PF1 within the price structure, structure and evolution of transformation costs, particularly the manufacturing overhead, and material costs for the last three years, and establishing the percentage of MCI target at process level, the evolution of customer demand and the accuracy of the forecasts, the evolution of product life cycles—planned vs. current, the state of development projects for new products and the MCI needs for MCI goal and MCI targets, etc.
■ *The analysis of all current OMIs and those required to be measured*: The calculation formulas, the units of measurement, the persons responsible for their continuous collection, the manager responsible for verifying their accuracy, their frequency of collection and the main challenges, the risks that may alter their accuracy, their evolution in time, establishing connections to detail levels—up to processes and activities (the initial state, the measurements performed by the MCPD team).

For PF3, the following measurements (OMIs) were targeted in particular: *market driven for MCI* (sales turnover and market share for target area), *profit driven for MCI* (manufacturing profit: unit profit, production capacity, production numbers, scrap ratio, rework for assembly), *MCI targets driven* (manufacturing unit cost: transformation unit costs and material unit costs), and *MCI means driven* (manufacturing lead time, production delivery performance, OTIF).

■ *Detailed analysis of KPIs for the manufacturing flow, detailing each OMI (stock of raw material, technological flowchart, and stock of finished products)*: The main purpose is to identify the difference between the current state of the manufacturing flow and its ideal state, by full application of the OPF principle, in order to identify subsequently the best opportunities for systematic and systemic improvement of processes at the shop floor level and beyond. Thus, analyses were performed for technological processes, process steps, process substages, technological operations, operation phases, parameters and principles of equipment operation, type and number of operators and supervisors, number of workstations, man*hours by product, total cycle time, overtime level, the specificity of bottleneck areas, maximum takt time, the level of balancing of operations and processes, current and required levels of synchronization for production takt time with supply activities and customer demand, OEE for six equipment (the synchronization level with takt time; accuracy of data collection), current capabilities vs. capabilities necessary in the future, distances on layout, average time and frequency for setup activities, times and routes for material handling, assembly line speed, WIP map, WIP behavior according to setup activities and process manipulations, production delivery performance (OTIF)—stock days of raw material and stock days of finished products, major and minor accident, occupational diseases, etc.

■ *Defining KPIs for losses and waste for each process, MKP* (Posteucă and Sakamoto, 2017, pp. 116–132): Calculating formulas, units of measurement, persons responsible for their continuous collection, the manager responsible for verifying their accuracy, their frequency of collection and the main challenges, the risks that may alter their accuracy, their evolution over time, the impact of current managerial decisions thereon (the way in which decisions to change the production plan or to allocate time for laboratory samples would be imposed by the quality assurance department), measuring the impact of processes and bottleneck activities on losses and waste (*critical losses and waste*) both at the level of the process in which the bottleneck is located, and especially along the entire manufacturing flow (downstream and upstream), establishing connections between *losses and waste* on levels of detail—processes and activities.

■ *Defining KPIs of CLW for each process, MCKP* (Posteucă and Sakamoto, 2017, pp. 132–140): Analysis of the current manufacturing cost-calculation system at product level, analysis of the current budget system (for each cost center, each equipment was declared as a cost center), the establishment of CLW calculation formulas and connections between *critical losses and waste and CLW* on levels of detail— processes and activities; CCLW determination.

■ *Analysis of the current state of implementation of systematic deployment projects (kaizen) and systemic deployment projects (kaikaku)*: Assessment of the status of each planned improvement project and analysis of delays in improvement projects, the reasons and impact of these delays in costs and profit; assessment of resource allocation methodology for improvements; assessment of current organization to facilitate people's participation in improvements; gains analysis for each project completed (time, production capacity, motion, saved and avoided costs, intangible effects, etc.).

Therefore, the MCPD team understood that in order to achieve a robust MCI policy deployment (targets and means), all manufacturing processes should allow a continuous knowledge of process-level CLW and manufacturing flow-level CCLW, as all critical KPIs of losses and waste are under control, to identify continuously the directions for improving CLW and CCLW, so that in time the MCI opportunities will decline and reach the *zero cost of losses and waste* status for as many parameters from as many processes as possible, if not all. Establishing MCI targets at all PF3 process levels should help to keep under control, reduce, and/or eliminate all conditions that lead to setting back from the *zero cost of losses and waste* state. The concept of *zero costs of waste and waste* refers to the fulfillment of the temporally set level for the MCI target and/or to the ideal state of lack of CLW for a certain process or of CCLW for all the manufacturing flow. The MCPD team has promoted the concept of *zero costs of losses and waste* throughout the plant, especially for the processes considered to generate critical losses and waste, and that implicitly spread CLW throughout and beyond the manufacturing flow.

MCI policy deployment involves two broad phases (see Table 1.3):

a. *Stage 1: Planning items to set MCI targets*
1. *Confirming the current state of the product family cost (PF3)*: The deployment starts from the OMIs for the *manufacturing costs* and continues at the seven levels of KPIs—starting from KPI1 and ending at KPI7. It was intended to confirm the standard cost level by comparing manufacturing standard costs with the current state (as measured by the MCPD team). Some inconsistencies (exceeding against the standard costs) have been identified at *transformation costs* (especially for manufacturing overhead: (1) maintenance costs, (2) utility costs, and (3) other manufacturing overhead, a lack of detail for about 35% of them), but also for *material costs* (raw materials costs and especially for auxiliary materials costs).
2. *Confirming the current situation of losses and waste for the product family (PF3) (concomitantly with the confirmation of the current situation of earlier mentioned costs)*: To support the subsequent CLW calculation at the process/work center/processes/ activities level (for the seven levels of KPIs for losses and waste— starting from KPI7 and ending with KPI1), each process was analyzed for PF3 as follows: setting checkpoints, average current value for a 30-day calendar interval, setting control method and frequency of data collection (manually or directly from equipment), and establishing the person responsible for data collection (usually operators); this way, KPIs were set for losses and waste.
3. *Identifying the critical losses and waste (weaknesses or manufacturing key points—MKP—which generates other losses and waste along the entire manufacturing flow and beyond)*: Critical phenomena from the processes have been identified that trigger the occurrence of other losses and waste, both in the process as well as throughout the entire manufacturing flow (upstream and downstream). An example of a *critical losses and waste* (MKP) phenomenon was the average breakdown time of 6 months for plastics painting equipment that had a major impact both on painting process times and on other processes (the causal relationship between breakdown losses, other generated losses and waste was determined).

Table 1.3 Example for MCI Policy Deployment

OMI	KPI1	KPI2	KPI3	KPI4	KPI5	KPI6	KPI7	Weaknesses	R	T	TT	MEANS 1: PST	MEANS 2: Kaizen (k)	MEANS 3: Kaikaku (K)	Means Impact	RM	MRT	MTT	Responsible Manager	Start–Finish	Expected Profit
Manufacturing cost on each product family or product																					
	Transformation costs								$X	$Y	%										
	Cost of losses for transformation costs								$X	$Y	%										
		Direct labor costs							$X	$Y	%										
		Cost of human work losses						○	$X	$Y	%	x	Assembly line balancing	X	○	$Z	$I	%	Max W.	1.03–1.06 (k)	$50,000
		Manufacturing overhead costs							$X	$Y	%										
		Cost of losses for manufacturing overhead costs							$X	$Y	%										
			Equipment manufacturing overhead costs						$X	$Y	%										
			Cost of losses for equipment manufacturing overhead						$X	$Y	%										
				Equipment scheduled downtime costs					$X	$Y	%										
				Cost of scheduled downtime losses				◇	$X	$Y	%	x	Reduce production stoppage due to lack of material	X	△	$Z	$I	%	John P.	1.01–1.04 (k)	$27,500
				Equipment effectiveness costs					$X	$Y	%										
				Cost of OEE losses					$X	$Y	%										

(Continued)

Table 1.3 (Continued) Example for MCI Policy Deployment

OMI	KPI1	KPI2	KPI3	KPI4	KPI5	KPI6	KPI7	Weaknesses	R	T	TT	MEANS 1: PST	MEANS 2: Kaizen (k)	MEANS 3: Kaikaku (K)	Means Impact	RM	MRT	MTT	Responsible Manager	Start–Finish	Expected Profit
					Equipment availability costs						%										
					Cost of availability losses				$X	$Y	%										
						Area 1…"n"			$X	$Y	%										
						Cost of availability losses for area 1…"n"			$X	$Y	%										
							Process 1…"n"	Θ	$X	$Y	%	x									
							Cost of availability losses for area 1…"n" in process 1…"n"		$X	$Y	%		Reduce breakdown	Repaint equipment	◉	$Z	$I	%	Tom I.	15.03–14.6 (k) and 20.02–30.11 (K)	$127,500
					Equipment performance costs						%										
					Cost of performance losses				$X	$Y	%										
						Area 1…"n"			$X	$Y	%										
						Cost of performance losses for area 1…"n"			$X	$Y	%										
							Process 1…"n"	Θ	$X	$Y	%	x									
							Cost of performance losses for area 1…"n" in process 1…"n"		$X	$Y	%		Improve initial setup in the plastic injection area	x	◉	$Z	$I	%	Stefan T.	20.02–19.05 (k)	$100,000
					Equipment quality costs				$X	$Y	%										
					Cost of quality losses				$X	$Y	%										

(Continued)

Table 1.3 (*Continued*) Example for MCI Policy Deployment

OMI	KPI1	KPI2	KPI3	KPI4	KPI5	KPI6	KPI7	Weaknesses	R	T	TT	MEANS 1: PST	MEANS 2: Kaizen (k)	MEANS 3: Kaikaku (k)	Means Impact	RM	MRT	MTT	Responsible Manager	Start–Finish	Expected Profit
						Area 1…"n"			$X	$Y	%										
						Cost of quality losses for area 1…"n"			$X	$Y	%										
							Process 1…"n"		$X	$Y	%										
							Cost of quality losses for area 1…"n" in process 1…"n"	Ǫ	$X	$Y	%	x	Reduction of scrap in the plastic injection area	X	○	$Z	$I	%	Tanya B.	30.06–30.09 (k)	$75,000
		Material and utility costs						◊	$X	$Y	%	x		Installation of measuring devices for "AA" class equipment	△	$Z	$I	%	James B.	15.05–14.08 (k)	$35,000
		Cost of material and utility losses							$X	$Y	%										
		Depreciation costs							$X	$Y	%										
		Cost of depreciation losses							$X	$Y	%										
	Stocks costs								$X	$Y	%										
	Cost of stocks waste							Ǫ	$X	$Y	%	x	Reducing WIP resulting from setup and moves	X	○	$Z	$I	%	Oliver S.	12.09–15.12 (k)	$85,000

4. *Identifying the current state of CLW for the product family (PF3)*: It was the first CLW calculation for the seven levels of KPIs—starting from KPI7 and ending with KPI1.

5. *Evaluating the process that generates CCLW for the product family (PF3)* (*MCKP*, which generates CCLW): By resuming the example mentioned earlier, CCLW was determined by summing the cost of breakdown losses for the painting process, all costs of losses and waste from the painting process, and costs of waste and waste along the entire manufacturing flow (upstream and downstream). One of the major challenges was that it has been identified that the current standard cost system does not accurately collect the specific consumption of auxiliary materials for each process. The MCPD team was very careful in this step as it was aware that future MCI targets will aim at MCKP, which generates CCLW, and MCI means targets will aim at MKP, which generates losses and waste.

6. Furthermore, each process has been evaluated from the perspective of its impact on *CCLW and CLW* as follows: Θ, high impact of process on CCLW and CLW (5 points); Q, some impact of process on CCLW and CLW (3 points); and Ǒ, limited impact of process on CCLW and CLW (1 point). Then the main *weaknesses that generated CCLW* were determined. It is about the 4Ms—machines, methods, materials, and manpower (people power)—but also about "mother nature" and measurements. For example, in *the technological flow of injected and painted parts*, in the *painting injected parts* phase, in the process of *preparing parts for painting*, 360 minutes of waiting per month for material or *internal logistics/handling losses* were measured in the first operation (*supplying workplace with parts to be painted*) (on average, for 24 hours there were three events of delay of the supply at the workplace by about four minutes each delay). The 360 minutes were considered as losses because they were not considered in the *time study* and implicitly in establishing the standard costs for *handling standard costs*. The costs of the 360 minutes related to the *internal logistics (handling) losses* were about $900 per month (in the structure of *cost of scheduled downtime losses*). The lost opportunity to perform manufacturing for the 360 minutes was also considered. But in the process of *preparing parts for painting, motion/walk losses* have also been identified for the *positioning parts on the painting frame* operation (times above

the standard ones already included in standard costs). The costs associated with time offsets on the *positioning parts on the painting frame* were about $350 per month. Therefore, *CCLW* for *preparing parts for the painting* process was about $1,250 per month (the first month when such measurements were made), and the process was marked with Θ (5 points) since it generated losses and waste along the entire manufacturing flow. Until then, this process was not deemed as a process to be considered in the improvement plans. Furthermore, it has been identified that the whole process of *internal logistics (handling)* causes an unpredictable level of losses along PF3 and needs to be improved. The *internal logistics (handling)* process was declared MCKP (*or CCLW*) as it caused other losses along the entire manufacturing flow for PF3 and not just $27,500 (forecasted for 12 months). The scoring system (5 points, 3 points, and 1 point) helped in directing subsequent MCI means.

7. *Establishing MCI targets*: Once the priorities for *CCLW* and *CLW* have been set and, implicitly, their current state (current state or baseline state; R, reference), the target level for MCI (T, targets) has been set, taking into account the cost decrease goal of 6% and meeting the MCI goal of $500,000 for the next 12 months. Normally, since this is the first MCI policy setting (targets and means), the bid to set MCI targets was very generous (much beyond the need for a unit cost reduction of 6%; in fact, it was about 30% in total; CLW were hidden in the structure of current and officially accepted standard costs that diluted target profits that could have been obtained). When establishing MCI targets, there were interdepartmental discussions on associated MCI means (through catchball process). Then, the mechanism for tracking the subsequent evolution of target achievement for MCI (TT, % of target touch; 25%, 50%, 75%, and 100%) was established until the end of the pre-established reference period (the next 12 months; the setting of MCI targets and means is a continuous activity of the managerial team with visibility for the next 12 months). In this context, a MCI target may have one or more MCI means for the next period (level 1, 2, or 3).

b. *Stage 2: Action items to set MCI means*

8. *Approaching deviations regarding CCLW and CLW from the predetermined acceptable value with the help of the Managerial Problem-Solving Techniques (means level 1, PST)*: To this end, the

following steps are pursued: The deviation from CCLW and CLW (the problem) is understood, the problem is broadly defined, the root cause is analyzed, alternative countermeasures are sought, each alternative (possible solution) is evaluated, one or more solutions are chosen, the solutions are implemented, and it is evaluated whether the deviation from the CCLW and CLW (the problem) was solved or not. If the problem has not been resolved, then the process is resumed until CCLW and CLW return to the target range, or a kaizen project (means level 2) is initiated. Typically, the A3 technique is used to address these issues. Under the pilot project, eight A3 projects were carried out to maintain CCLW. One of the challenges of this step was the establishment of the escalation system and the necessary actions to be carried out at each hierarchical level—a challenge subsequently addressed through the daily MCI management.

9. *Reducing or eliminating CCLW and CLW with the help of systematic improvement projects (means level 2, kaizen—k)*: Like the MCPD system and MCI policy deployment, kaizen projects follow the PDCA cycle. In order to achieve the *zero costs of losses and waste* state for as many parameters as possible, from as many processes as possible, if not all, it was necessary for all people in PF3 and many of AA-Plant to be involved, not just in the manufacturing area. Six kaizen teams were formed for the six opportunities identified to meet the MCI targets: (1) *assembly line balancing*, (2) *reduce production stoppage due to lack of material*, (3) *reduction breakdown*, (4) *improve setup time in the plastic injection area*, (5) *reduction of scrap in the plastic injection area*, and (6) *reducing WIP resulting from setup and transfer*.

10. *Eliminating CCLW and CLW with the help of systemic improvement projects (mean level 3, kaikaku—K)*: If *CCLW* and CLW cannot be managed by *means level 1* or reduced by *means level 2*, then *means level 3* or projects of radical/systemic improvement (kaikaku) are suggested, because the improvement or elimination of *CCLW* and CLW must be made no matter what implications can be generated. The decision to address *CCLW* and *CLW* through kaikaku projects always considered the need for future capacity—imposed by customers. At PF3, a team was set up for a kaikaku project to *replace painting equipment*.

11. *Assessing the means impact*: It is performed to determine which means are priorities and how to allocate the necessary resources:
    - ■ ⊚ with 5 points, high impact of MCI means on MCI targets in the process
    - ■ ○ with 3 points, appropriate impact of MCI means on MCI targets in the process
    - ■ △ with 1 point, small impact of MCI means on MCI targets in the process

12. *Establishing MCI means targets*: Once the priorities for means impacting on *CCLW* and *CLW* have been set and, implicitly, their current state (MR, means reference), the target level for MCI means (MRT, Means-Reduction Target) has been set. An example of MCI means target was for a setup target which, by doing so, will cause reaching the MCI target for setup at the expected level, contributing to achieving the unit manufacturing cost. Thus, it was set to reduce the 45 minutes of setup time to *20 minutes to meet the MCI targets related to CCLW in the plastic injection process*. Again, the goal of reducing the cost of 6% and meeting the MCI goal of $500,000 for the next 12 months for all Means-Reduction Target was taken into account. Then, it was set to track the subsequent evolution of reaching the target set for MCI means (MTT, % of means target touch) by the end of the pre-established reference period (the next 12 months). This information was communicated to all those involved in the PF3 activities, especially to the operators involved in the improvement projects, on the *special panel dedicated to MCPD at the information center.*

13. *Project management and evaluation of expected results*: For this purpose, the following have been set up: *responsible manager* for each improvement project (means levels 1 and 2) or problem solving (means level 1), *start and completion date* of each improvement project (especially for means levels 2 and 3) or problem solving (means level 1), and the *necessary resources available.*

14. *Information centers*: The following have been developed for MCI management—*company action board, shop floor action board* (*with daily management action board*), and *small group action board*. The purpose of the information centers is to provide the necessary information in time to establish MCI targets and means. The mechanism for daily MCI management has been created for operational performance to be measured at each company level

(CLW and CCLW) and with a clear definition of the decision-making scope at each level. Vertical and horizontal communication flows and alert rules were designed to provide a quick response to MCI targets and means. At the same time, information centers have been presented (broadly) with all ongoing improvement projects and completed projects (briefly) to achieve MCI targets. Continuous updating with data and information, choosing priority setting criteria, escalating procedure, and monitoring the confirmation of reaching MCI targets over time have been the main challenges of the pilot project in this step. To begin with, the management team considered that the information on the MCI goal of $500,000 and the 6%-unit manufacturing cost reduction target should not be displayed at the *information center*. Both the target profit of $500,000 and the 6%-unit cost reduction goal were fulfilled successfully by meeting MCI targets and means.

As a general conclusion, MCI policy deployment considers continuously the need to achieve target price and, implicitly, target cost and target profit obtained from improvements. Often, some companies, even if they follow good directions in general, have difficulties in meeting these directions. So, in the case of MCI policy deployment too, *MCI means represents the core area of MCI policy deployment*. Over time, MCI policy deployment will generate an evaluation of all managers by comparing the level of *MCI goal* to the level achieved. In this context, MCI policy deployment underpins the development of *annual improvement and cash budgets for existing and new products*, which is the quantifiable performance level required to be fulfilled in accordance with the long-term objective of reducing manufacturing unit costs and the basis for the multiannual plan of productivity. The basic logic of MCI targets and means is to establish MCI targets and means at the level of every process of PFC and/or every product starting from the current level of CCLW and *CLW* and by connecting this level to the need to reduce the price/cost imposed by the competitiveness level on the market. Sometimes, incremental MCI targets and means can be used that do not take much into account the exact need to reduce unit manufacturing costs imposed by the market. Therefore, monitoring the evolution over time of MCI targets and means in the MCPD system logic to support a long-term target profit is achieved with the help of *annual improvement and cash budgets for existing and new products* as an *integral part of the annual master budget* (Posteucă and Sakamoto, 2017, pp. 162–193).

1.6 Stakes of MCPD: Uncovering Hidden Reserves of Profitability

The long-term MCPD system's stake is to achieve the internal profit level to meet the target profit or the multiyear profit plan at the company level, even if sales volumes fluctuate (see Formula 1.1). On an annual basis, the MCPD system stake is to achieve the annual MCI goal group-wide (see Formula 1.3), to achieve annual target profit (see Formula 1.2), and to achieve MCI goal deployment at each plant level in the group (see Formula 1.4) at the level of each process of each PFC (see Formula 1.5), by setting MCI targets for each process for CLW and CCLW (see Formula 1.6) and by establishing MCI means for each process by setting targets for losses and waste (see Formulas 1.8 and 1.9).

To support the MCI goal deployment achieved by establishing MCI targets and means scientifically at the level of each PFC process, it is determined from the beginning the MCI level needed to be achieved for each product or each important product of each PFC to achieve the level of the target profit from sales (external profit) and from the reduction of non-value-added costs (through MCI) (see Formula 1.7).

Achieving MCI goal deployment across the entire company will result in an acceptable return on investment (return on investment = net income/investment; net income = gross profit – expenses) on the basis of gross profit growth (achieving the target profit) and reduction of expenses (increase of internal profit on the background of annual MCI goal fulfillment). This way it is possible to avoid unnecessary investments by increasing the productivity of current capacities and implicitly by increasing the competitiveness through profit.

Meeting the annual MCI goal at the level of each PFC of a plant will fill the "empty space" in the annual target profit so that the company can have an acceptable and sustainable development. This way, unnecessary costs are saved from the unit manufacturing cost structure, which leads to an increase in competitiveness by price.

Going now to the manufacturing flow level, the main stake of the MCPD system is to achieve a continuous and harmonious transformation of all manufacturing processes by continuously connecting to the need to reduce the unit manufacturing cost to uncover hidden reserves of profitability through the two types of reconciliations: (1) between profit driven for MCI/annual MCI goal and MCI targets and (2) between MCI targets and MCI means through catchball process at all organizational levels. This way, the pressure to reduce unit manufacturing costs and to obtain target profit for each PFC is transposed in an acceptable timely manner to the processes and activities level (top-down approach or the request for MCI targets), and then an answer is received in an acceptable time (bottom-up approach, or offer for MCI means and implicitly for MCI targets).

To this end, locating, collecting, and quantifying CLW and CCLW are the elements that help establish MCI targets deployment by analyzing and interpreting manufacturing costs and losses and waste for each area and every process of each PFC, to evaluate, plan, and control all MCI means in order to continuously reduce manufacturing costs at all cost centers, to continuously reduce unit manufacturing costs, and to continuously uncover hidden reserves of profitability.

The stake concept is preferred to the concepts of benefits and/or advantages since it is more comprehensive and better captures that the MCPD system is a system continuously connected to the market (customers, competitors, and suppliers), to the competition, and market play or numerous pressures of the market are continuously present in the decisions required to be taken in the MCPD system. The pressures to meet cost reductions in a predetermined time frame by continuously presenting a more generous offer for MCI (CCLW and CLW) make the MCPD system become a stake by itself in the *competitiveness through price and profit* game. Therefore, the responsible use of all resources in the manufacturing area through the MCPD system has external and internal stakes within a manufacturing company.

The major external stakes of the MCPD system are as follows:

1. *Continuous support of price/cost competitiveness at the level of each PFC and/or product* (flexibility by cost).
2. *Continuous support of the long-term profit plan (multiannual target profit)* by continuously providing an acceptable level of manufacturing profit obtained by continuously identifying and exploiting *CCLW* and *CLW* with the help of productivity increase. The continuous exploitation of *CCLW* and *CLW* through setting MCI targets and means at PFC process level generates both external manufacturing profit by increasing capacity (especially in the case of increased customer demand—sales volumes) and internal manufacturing profit by continuously reducing or eliminating unnecessary consumption (especially in the case of reducing customer demand—sales volumes).

The structure of *CCLW* and *CLW* covers a part of the accepted current costs that dilute a part of the manufacturing target profit that could be obtained but which is wasted.

3. *Reducing investment by increasing the use of current capacities—* especially for equipment.

The main internal stakes of the MCPD system are as follows:

1. *The continuous targeting of all productivity improvements through the need for cost reduction* imposed by customers (without affecting the quality level or delivery times) in order to maximize the output level and to minimize the use of inputs determines the following tangible effects (continuous achievement of targets for OMIs): manufacturing profit, manufacturing costs, total lead time, sales factory turnover, manufacturing number, manufacturing capacity, market share, scrap and rework ratio, man*hour by products, kaizen benefits, stocks days for raw materials and finished products, major accidents, etc. Furthermore, remarkable results are achieved on self-financing capacity steadily increasing, continuous cash flow improvement, and easier access to external financing.

2. *Exact allocation of resources for improvements* that are required for a continuous transformation of the manufacturing companies to meet the production capacity target in terms of efficiency, knowing in advance, on a continuous basis, the possible level of the manufacturing target profit, behind *CCLW and CLW,* and the concrete possibilities of reducing the unit manufacturing costs through productivity. For this purpose, the performance level of MCI policy deployment is continuously pursued with the help of *annual improvement and cash budgets for existing and new products.*

3. *Continuous development of scenarios for MCI targets and means (by simulations),* by continuously involving all people at all levels to mitigate pressures on the level of costs between top-down (the need of cost reduction originating from the market) and bottom-up (the MCI offer originating from the processes of each PFC), determines the continuous improvement of teamwork among all employees, work satisfaction, knowledge of operators, and dealing with problems systematically (means levels 1, 2, and 3). In this way, CCLW and CLW continuously drive productivity improvement toward profit. *CCLW is in fact critical costs of non-productivity.*

4. *The desirable contextual behavioral identity of managers* is continuously improving, which creates the premise for a pro-creativity and pro-productivity environment from the management perspective (*management branding*, MB) (Posteucă, 2011). Coordinated by the continuous offering of concrete directions for obtaining manufacturing target profit and reducing unit manufacturing costs, directly from the internal processes of the manufacturing company, the desirable behaviors of managers improve over time.

5. *Daily cost management* (DCM) (means level 1) *becomes a lifestyle at the shop floor level*, and the operators' confidence in themselves and in the company is continuously improving.

6. *Continuously ensuring an acceptable level of unit manufacturing costs for new products* to be launched and requiring cost reduction (monitored through *annual improvement budgets for new products*). Achieving the profit plan of new products throughout their life cycle is continuously sustained through the MCPD system.

Therefore, in light of the call of Peter F. Drucker in 1963 on the need to develop a managerial tool to balance the relationship between efficacy and efficiency ("What we need is (1) a way to identify the areas of effectiveness (of possible significant results), and (2) a method for concentrating on them") (Drucker, 1963), and the fact that "Efficiency is doing things right; effectiveness is doing the right things" (Drucker, 1973), and Ohno's definition (Ohno, 1988, p. 8) of efficiency ("Efficiency, in modern industry and business in general, means cost reduction"), the successful completion of these stakes creates the framework for a continuous and harmonious transformation of manufacturing companies, transformation coordinated by the need for cost reduction and obtaining manufacturing profit imposed by the market, and by increasing the productivity imposed by the company in response to market needs. Thus, the MCPD system through MCI policy deployment makes the cost the turnkey between market needs (low cost and profit) and internal needs to improve company productivity, and the self-pervading effects of uncovering hidden reserves of profitability come from (1) increased use of current and future capabilities and (2) MCI with impact on the continuous reduction of unit manufacturing costs.

References

Akao, Y., 1991. *Hoshin Kanri: Policy Deployment for Successful TQM* (originally published as Hoshin Kanri Kaysuyo no jissai, 1988). New York: Productivity Press.

Drucker, P., 1973. *Management: Tasks, Responsibilities, Practices.* New York: Harper & Row.

Drucker, P. F., 1963. *Managing for Business Effectiveness.* Boston, MA: Harvard University, Graduate School of Business Administration.

Hirano, H., 2009. *JIT Implementation Manual—The Complete Guide to Just-In-Time Manufacturing.* Hoboken, NJ: CRC Press.

Imai, M., 1997. *Gemba Kaizen: A Commonsense Low-Cost Approach to Management.* New York: McGraw-Hill.

Nakajima, S., 1988. *Introduction to TPM: Total Productive Maintenance.* Cambridge, MA: Productivity Press.

Ohno, T., 1988. *Toyota Production System: Beyond Large-Scale Production.* Cambridge, MA: Productivity Press.

Posteucă, A., 2011. Management branding (MB): Performance improvement through contextual managerial behavior development. *International Journal of Productivity and Performance Management*, 60(5), 529–543.

Posteucă, A., 2015. Manufacturing cost policy deployment by systematic and systemic improvement. PhD dissertation, University Politehnica, Bucharest, Romania.

Posteucă, A. and Sakamoto, S., 2017. *Manufacturing Cost Policy Deployment (MCPD) and Methods Design Concept (MDC): The Path to Competitiveness.* New York: Taylor & Francis.

Shingo, S., 1989. *A Study of the Toyota Production System: From an Industrial Engineering Viewpoint.* New York: Productivity Press.

Shirose, K., 1999. *TPM: Total Productive Maintenance: New Implementation Program in Fabrication and Assembly Industries.* Tokyo, Japan: JIPM.

Chapter 2

MCPD System: Overview and Dynamics of Business Contexts

The main purpose of the Manufacturing Cost Policy Deployment (MCPD) system is to translate the long-term cost reduction strategy into actions and activities at each hierarchical level and at the level of every person in the company. In summary, in order to meet the long-term cost reduction strategy and, implicitly, the multiannual target profits, especially internal target profits, at plant level, the MCPD system represents the following:

- A trajectory of annual actions that manufacturing companies plan to achieve the MCI goal at the level of each product family cost (PFC) process [*manufacturing cost improvement (MCI) targets deployment*]
- Gradually organizing the entire company at the level of each process of each PFC in order to determine the role of each current and/or future resource in fulfilling MCI means by continuously implementing the solution of systematic (kaizen) and systemic (kaikaku) improvement projects (*MCI means deployment*)

The MCPD system is applied at the top of a manufacturing company and then deployed down through each department, section, or process for each PFC individually, or it can be applied *stand-alone* within a department or just a PFC requiring a significant cost reduction. MCPD principles, phases, steps, and judgments are the same in both situations. Normally, through continuous use, the MCPD system improves year by year.

In this chapter, the theoretical MCPD system and its approaches in the medium and long terms are presented. In Section 2.1, the transition from productivity vision to the annual actions plans for each PFC is presented. Then, in Section 2.2, the benefits of the MCPD system are presented for each hierarchical level in the company. In Section 2.3, there are presented the aims and approaches of the three phases and seven steps of the MCPD system. In Section 2.4, the main ingredients of the long-term manufacturing target profit are presented from the perspective of the MCPD system, namely price benchmark analysis, synchronization of life cycle targets (profit, price, sales, capacity, productivity, and cost) of investment and productivity. Finally, in Section 2.5, the strategic alignment of opportunities for improvement to the MCI need is presented.

2.1 From Productivity Vision to Actions Plans for Each PFC

The systematic MCPD planning process used to meet the need to reduce manufacturing costs and the multiannual profit plan with the daily activities of all people in and beyond the company ensures that everyone is aware of the vision and mission of productivity and the daily role to meet the MCI targets to achieve the annual MCI goal by increasing productivity. The basic concept of MCPD is relatively simple: if all the company's resources are directed to meet the annual MCI goal, this reduces the unit manufacturing costs by increasing productivity; thus the company moves in the right direction and leaves the competitors behind. If some departmental processes, or even an activity, are not in the right direction (at PFC level), then we have to stop and look for solutions to reorient them to the preset direction (MCI targets). At the same time, if the direction is correct, then all the current practices will be critically and constructively analyzed to turn the continuous improvement of manufacturing costs into a way of life at all levels of the company.

The *productivity vision* is the concern of the board of directors. They are constantly concerned about the direction the company must go to achieve target profits for the next few years (usually for the next five years; the long-term profit plan). Specifically, *What types of products? What quantities of products? And on what markets could these be sold?* The board of directors is continually rethinking such questions in an articulated form about every four months. Further on, starting from productivity vision, the senior managers develop the *productivity mission* every two months. Their primary purpose is to determine the level of capacity to be assured to support the sales volumes alleged and assumed by the board of directors (usually for the next minimum three years). In fact, it is about understanding the current external (especially by the board of directors) and internal (especially senior managers) situation. Both the board of directors and senior managers need to understand exactly where the company lies and what will influence the company significantly in the coming period from the perspective of achieving the long-term profit targets. To this end, external and internal strategies are assessed, especially the internal cost reduction strategy; external and internal key factors are defined; and current performance levels are assessed against some values considered benchmarks.

Furthermore, senior managers together with middle managers set out for each PFC:

1. *MCI targets and means*
   - *Market driven for MCI*: target sales turnover, target production numbers, and target price (average)
   - *Profit driven for MCI*: target profit, target external profit, target internal profit; annual MCI goal; annual MCI goal for each PFC; annual MCI targets for processes of each PFC, number of *kaizen* projects for MCI targets in the all departments; number of *kaikaku* projects for MCI targets; target profit from new products
   - *Driven by MCI targets deployment*: unit manufacturing costs (the cost per minute for each process work center/product; material costs for each product and for each PFC); percent of value-added manufacturing costs from total manufacturing costs; percent of non-value-added manufacturing costs from total manufacturing costs; costs of losses and waste (CLW); total costs of losses and waste (TCLW); critical costs of losses and waste (CCLW), total critical costs of losses and waste (TCCLW); annual MCI targets on each process/ work center/product (per each type of CCLW and CLW); annual MCI targets for non-value-added manufacturing cost per unit (for transformation costs and for material costs). Continuous reconciliation of MCI targets with MCI means with catchball process
   - *Driven by MCI means deployment*: KPI targets for losses and waste for each process of PFC; total losses and waste (TLW), critical losses and waste (cLW)—based on the causal relationships along the

manufacturing flow; total critical losses and waste (TcLW); MCI means targets deployment for all the processes of each PFC, levels 1, 2, and 3

2. How to develop *the annual improvement budgets* for each PFC, both for existing and new products in order to support the MCI on the short, medium, and long term by providing a clear visibility of the current state of MCI targets

3. How to stratify the strategic cost reduction intent by MCI at the level of the key processes and activities (CCLW and CLW) of each PFC (1, the basic cost strategy; 2, the interdepartmental cost strategies; and 3, the PFC strategies)

4. The development and monitoring mechanisms of the *annual action plan for MCI* for each PFC (start and finish dates of plans for systematic improvement activities, kaizen for MCI, and systemic improvement actions, kaikaku for MCI); timely allocation of all necessary resources (allocation of the time needed for people to participate in the improvement projects and the necessary material resources) in order to fulfill on time and accurately the targets for MCI means.

5. How to *engage the workforce*, departmental organization, ownership for MCI targets and means for each process/work center/product of PFC and resources for determining training needs related to the fulfillment of the MCI targets and means; the role of each employee in achieving the MCI targets and means of the company and implicitly a clear visibility on how to set MCI targets and means for each PFC; bidirectional two-way communication mechanisms for MCPD, up and down, to establish MCI policy (by catchball process); ownership for MCI targets and means for each process/work center/product of each PFC

6. The way to continuously monitor the tangible and intangible effects of *MCI performance management* at company level and beyond

7. The mechanisms of the *daily MCI management* to monitor critical processes and improve performances by developing contextual managerial behaviors, in line with employee expectations (management branding) (Posteucă, 2011; Posteucă and Sakamoto, 2017, pp. 240–244)

For the success of MCPD, besides MCI means deployment (level 2, kaizen) at the level of processes and activities of each PFC, to achieve continued improvement of CCLW and CLW, plans for MCI means deployment (level 3, kaikaku) are established for the following:

■ *New management methods* (system approaches, such as the implementation of *knowledge management, autonomous maintenance, cybernetic breakdown management,* or specific elements of

industry 4.0: big data and analytics; autonomous robots; simulation; horizontal and vertical system integration; the industrial Internet of Things; cybersecurity; the cloud; additive manufacturing, augmented reality, etc.)
■ *Individual plans to improve people's knowledge and skills*

Consequently, all activity plans and actions to meet MCI means targets must be converging to MCI targets, which, in turn, converge to the annual MCI goal and then to the mission and vision of productivity and profitability (long-term profit/ targets profit). This ensures the development of the productivity master plan (one, three, and five years) that must provide continuous feedback on the level of performance achieved of the productivity and profitability at each PFC level.

2.2 MCPD System: The Benefits for Each Hierarchy in the Company

Productivity vision deploying and the continuous detailing of the company's intentions at all levels is a continuous necessity for manufacturing companies. From the perspective of the MCPD system, such intentions (starting from productivity vision deploying) refer especially to (1) product number increase, (2) cost decrease, and (3) quality increase. Furthermore, in order to meet the intention to increase the number of products (output maximization), there is a need for productivity increase, product range increase, and capacity increase projects. To satisfy the intent of cost decrease (minimizing inputs), there is a need for decreasing material costs, decreasing depreciation costs, decreasing design costs, and decreasing transformation costs. To meet the intent of quality increase (maximizing outputs), there is a need of increasing customer satisfaction and increasing quality product ratio.

Decreasing material costs and especially transformation costs are the main focus areas of the MCPD system and, implicitly, MCI targets and means. For this, all levels of the company are involved, just like for all other business dimensions:

1. *Senior-level managers* (or top-level managers, such as senior or board executive, functional heads, managing director, and director or general manager) who make strategic decisions on the reduction strategy of the manufacturing costs (achieving the internal profit) and its contribution to the long-term profit plan (achievement of the target profits), both through the concrete reduction of unit manufacturing costs (minimizing inputs at the level of man/woman, machine, material, money;

achievement of the annual MCI goal) and by increasing sales volumes supported by a competitive price/cost and an acceptable level of quality (realization of MCI means targets)

2. *Middle-level managers* (such as department manager and section manager) who make tactical decisions on transposing the manufacturing costs reduction strategy (the achievement of internal profit) in the everyday activities of all employees in the company (MCI means)

3. *Lower-level managers* (such as subsection manager, supervisor, and foreman) who have the task of supervising the pursuit of daily activities for maintaining and continuously improving the manufacturing costs, especially the transformation costs

4. *Shop floor level* (such as *staff, team leader,* and *worker*) which must internalize the *zero costs of losses and waste* status, as well as anyone else, both within and beyond the company, achieved both during problem-solving projects [problem-solving techniques (A3 in particular)] and improvement projects (kaizen in particular, but also kaikaku) by ensuring continuously updated internal communication (updating the information center with data and information on the MCPD system)

In particular, the benefits of the MCPD system for each hierarchical level in the company are as follows:

■ *For senior-level managers*: Coordinated by the productivity vision (the continuous setting of multiannual target sales volumes to ensure the target profit), implicitly by the need to achieve internal profit, which has an impact on the increase in capacity and current quality, senior managers act as *a real strategic internal profit team*. The long-term profit plan (target profit), the external profit plan, the internal profit plan, the annual target profit, the annual external profit, and especially the annual MCI goal underlie each of their key decisions. Thus, the planning of the annual target profit, the annual external profit, and the annual MCI goal takes place transparently, making everyone involved with the company so easy to see and understand what is going on in the future. The annual external profit is the result of the expected annual profit on sales based on yearly budget development. If the annual external profit is not sufficient, then the annual MCI goal level is established (for each PFC and which is accomplished by reaching MCI targets and MCI means targets). This promotes the development of the organizational culture directed toward objectives or *cross-functional approach* and deviating from functional thinking (fulfilling MCI targets and means becomes a priority through the full involvement of all departments and people in and beyond the company). At the same time, the MCPD system, through MCI policy deployment, provides a unique way of translating productivity vision into productivity mission, in productivity and cost strategies, and in setting daily tasks for every person from the perspective of continually reducing the manufacturing costs. This unique annual way of the company to participate in the multiannual profit plan (target profits) ensures ownership and accountability that is directed top-down for each PFC. The involvement of senior managers from the very beginning in PFC logic and MCI targets and means to meet the multiannual profit plan and the annual MCI goal ensures the allocation of resources focused on reducing CCLW and CLW (understanding the needs for capacity growth at the level of important processes, a need to be met by kaizen and/or kaikaku) and, implicitly, on the performance through *annual improvement and cash budgets for existing and new products*. This ensures visibility and acceptable engagement at the next level down through *catchball process*. The basic question of the senior managers is: *What should be done in the company to meet the mission of the company's productivity?* In conclusion, with the help of MCPD, the senior manager will be able to answer two questions on an annual basis: *What percentage is required to be met by the annual MCI goal to achieve annual target profit? What contribution should each PFC bring to the annual MCI goal?* Senior managers will negotiate answers to these two questions with middle-level managers. At the same time, they will check periodically the reach of MCI targets and their performance on planning to achieve the annual MCI goal (twice a year: in July and in January).

■ *For middle-level managers*: These, coordinated by the productivity mission, will be able to clarify how synchronization of all departments is accomplished to meet the annual MCI goal through the deployment of the interdepartmental improvement projects (MCI means). Continuous awareness of *CCLW and CLW* at the level of each PFC process, coupled with productivity mission, provides clear visibility of the annual and multiannual managerial priorities and a detailed access to data and information. With the help of the MCPD system, one can better determine the personal development needs of middle managers, and not only that, from the MCPD system perspective, any training program needs to be converged with reducing and/or eliminating at least a *CCLW* or a *CLW*. This way, the feasibility of the training programs becomes easier to determine as the stake of these programs is known at the level of MCI targets. Moreover, in addition to continually determining the professional development opportunities of middle managers, their ability to continually discuss, within the *MCI catchball process*, with both senior managers and operators about the MCI targets and means levels for the processes coordinated by them, creates the premises of a creative and stress-free work environment based on real measurements and facts. This way, middle managers can constantly challenge MCI targets and means, but without seeking culprits and without creating unnecessary tensions in the company. The basic question of middle managers is: *How will the annual MCI goal be achieved?* In conclusion, with the help of MCPD, the middle manager will be able to answer three questions continuously: *How will the MCI targets be established? How will the MCI targets be met? What resources are needed to be allocated and when, in order not to waste potential opportunities for price competitiveness?* Middle managers will negotiate the answers to these questions with lower-level managers and MCPD implementation teams.

■ *For lower-level managers*: The greatest benefit to them is that they always understand their place and role within the company and what is expected from them. At the same time, evaluations of their participation in meeting MCI targets are made objectively and quantitatively. With the help of the MCPD system, any lower-level managers can see and understand in detail the current state of the company at its level of activity.

■ *For shop floor level*: With the help of the MCPD system, especially by continually seeking MCI means, a sense of pride and ownership is encouraged within the company, which leads to the concentration of all resources on MCI means and reducing and/or stopping unnecessary activities.

Therefore, the MCPD system is a critical success factor for manufacturing companies at all hierarchical levels. Without the implementation of this system, manufacturing companies may suffer from a lack of direction in terms of reducing unit manufacturing costs, and cost reductions can only come as a result of kaizen and kaikaku projects, not as a previously well-defined goal. In this context, where the unit manufacturing cost does not benefit from a rigorous planning of their reduction, as is the case with the MCPD system, the profitability of the manufacturing company is left in the hands of the competitors, customers, and suppliers in a fairly large proportion.

2.3 Three Phases and Seven Steps of the MCPD System

The MCPD system provides all the necessary steps for planning, implementing, and evaluating the performance obtained to achieve the harmonious transformation of manufacturing companies from the perspective of the need to continuously reduce costs, following the plan–do–check–act cycle, or the Deming cycle. More specifically, the MCPD system addresses the shift toward the betterment of CCLW and CLW in a planned way, or rather the cost of non-productivity. MCPD is a system because it represents a set of processes designed to meet MCI targets and means consistently coordinated by the need for cost reduction imposed by the market.

The MCPD system is structured in three phases and seven main steps aimed at translating objectives and cost reduction strategies into annual and daily activities and actions.

Figure 2.1 shows the key details of the phases and steps that are necessary to effectively implement the MCPD system:

- ■ *Phase 1: Manufacturing cost policy analysis.* Aims to provide structured planning that combines the need for cost reduction for each PFC (annual MCI goal) with the possibilities to meet this CCLW and CLW need in processes (annual offer for MCI targets to reach the annual MCI target), depending on the current possibilities of MCI means:
  - *Step 1: Context and purpose (Plan).* Aims at measuring and understanding as accurately as possible the external and internal context of critical points of a manufacturing company within a *productivity business model (PBM)* (Posteucă and Sakamoto, 2017, p. 21). For this, some activities are carried out, such as establishing the annual need for internal profit (annual MCI goal), price benchmark analysis, analysis of the structure and price evolution and the manufacturing costs, measuring the manufacturing company as a whole, measuring production flow from the perspective of non-productivity (losses and waste), defining manufacturing flow in ideal state for each PFC, measurement of CCLW and CLW for each PFC.
  - *Step 2: MCI targets and means (Plan).* Aim to establish MCI policy deployment (targets and means) for the process for each PFC in order to meet the annual MCI goal, regardless of whether sales are rising or falling. The annual MCI target setting for each PFC is achieved through the *external reconciliation* (originating from outside of the company) between the target price and annual manufacturing target

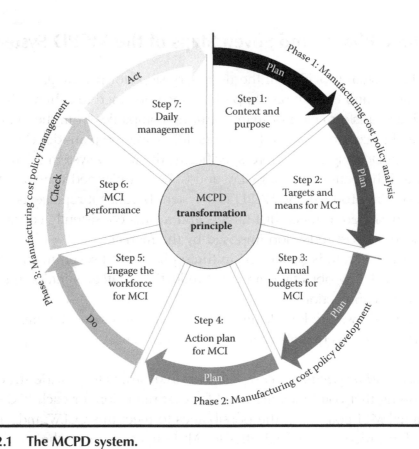

Figure 2.1   The MCPD system.

profit and the *internal reconciliation* between annual manufacturing target profit and annual MCI targets and means. Establishing MCI targets at the level of PFC processes is the key point of the MCPD system for achieving the annual MCI goal, and this is achieved by involving all employees in and beyond the company by searching the best way available at that time to meet MCI means targets (through catchball process). Approximately 70%–80% of the total annual available time of senior managers is for setting MCI targets and means to meet the annual target profit (especially annual MCI goal).

■ *Phase 2: Manufacturing cost policy development.* Aims at developing the systems and processes needed to translate MCI targets and means into consistent improvement activities targeted continuously at achieving the vision, mission, strategies, and productivity goals.

– *Step 3: Annual improvement and cash budgets for existing and new products (Plan).* The purpose of developing such budgets is to precisely plan the level of performance required for MCI by implementing solutions discovered through improvement projects. This ensures the long-term sustainability of the MCPD system. *Annual manufacturing improvement budget for existing products* aims at setting MCI targets at each manufacturing cost structure in each manufacturing process of each PFC. *Multiannual manufacturing improvement budget for new products* aims at identifying future opportunities to improve manufacturing costs at least in the first months after launching new products, but also throughout their life cycle. *Annual manufacturing cash improvement budget* aims to monitor the performances of projects to improve manufacturing costs in terms of cash flows generated.

– *Step 4: Action plan for MCI for each cost product family (Plan).* Performing the planning of *MCI means* (systematic and systemic improvement; kaizen and kaikaku) to ensure that all the resources needed to meet MCI targets and other business goals (three to five goals in total; to obtain the annual external profit) are available exactly when they are needed. At the same time, an individual plan for participation in MCI is established. Approximately 7%–8% of the total daily availability of operators is allocated to MCI activities (especially for MCI means level 2, kaizen).

■ *Phase 3: Manufacturing cost policy management (Do–Check–Act).* Aims at the full involvement of all the people in the company, and beyond, to fulfill the MCI means, monitoring, assessing, and adjusting the effectiveness of MCPD.

– *Step 5: Engage the workforce to execute the MCI targets (Do).* It is performed by organizing each department from the perspective of fulfilling MCI means targets, at the manufacturing flow level of each PFC by carrying out improvement projects and implementing the solutions identified, by precisely establishing training needs to meet MCI means, and by establishing the initial and updated annual training plan to achieve MCI targets and means.

– *Step 6: MCI performance management (Check).* The current performance is measured for: (1) the annual MCI goal, (2) the annual MCI targets and means, (3) the annual improvement budgets for each PFC, (4) the resources consumed for improvements against the benefits realized and planned, (5) the level of employee involvement in achieving MCI targets and means, and (6) tangible and intangible effects of improvement projects for MCI.

– *Step 7: Daily cost management (Act).* Aims to routinely monitor the fundamental aspects of operations from the perspective of meeting MCI targets and means and beyond, by daily observing critical points that generate CCLW and CLW; developing the contextual desirable managerial identity to meet MCI targets and means (*management branding, MB*), by addressing any deviation from MCI targets by running projects using different problem-solving techniques (especially A3) and by performing daily sessions structured at the shop floor level.

The implementation of the three phases and the seven steps can be accomplished at the same time or separately, depending on the maturity of the culture of continuous improvement of the company, and according to the experience, knowledge, and involvement of senior managers.

2.4 Long-Term Target Profit: Main Ingredients

The first element of external reconciliation is between target price and annual manufacturing target profit. Continuous knowledge of current and/or potential pressure on unit manufacturing costs from the perspective of target price and targets profit represents a desideratum irrespective of whether the manufacturing company delivers finished products directly to the final consumer or not. Annual target profit would be easier to achieve if sale prices could increase. However, in a highly competitive and rules-based-on-the-supply-and-demand principle market, with no dumping practices, sale prices have a limited range of impact, better known as growth in a company.

2.4.1 Price Benchmark Analysis

Price benchmark analysis is designed to substantiate strategic decisions that aim at maximizing external profit by determining the optimum volumes of products manufactured and sold, by setting the level of expected revenue and associated costs. Under normal circumstances, the price of a product and/or service is the result of a play between supply and demand, involving customers, competitors, and the company that provides the products and/or services at the level of expected costs and profits. The price benchmark analysis stake is to determine a long-term competitive price behavior of a company by analyzing the past, present, and, eventually, future actions of competitors and customers.

2.4.1.1 Identifying Comparative Companies

Competitive price behavior is influenced by the level of price pressure imposed by the competitor and customers. As is well known, both in the short term and long term, in highly competitive markets, companies have no control over prices and are forced to accept the price level imposed by the multitude of competitors or the power exerted by some few customers. In this context, in order to maximize long-term profit, the key dimensions are: production volume, productivity and continuous cost reduction. Any increase in sales price may lead to increased stock levels, reduced sales volumes, and, implicitly, reduced target profit. In contrast, in markets with lower competitiveness, the value perceived by customers for products and/or services rendered and the purchasing power are those that determine the price level and, implicitly, the cost structure.

The MCPD system focuses on the issues of markets with high competitiveness and with constant pressure on prices and cost level. Starting from Ohno's Toyota cost principle (Ohno, 1988, p. 8) (sale price – profit = costs, respectively), two main questions arise: *What are the most important customers and/or competitors that exert the greatest pressure on the costs of existing and future products?* and *How can we find out/have an insight into, on a continuous basis, the cost performance of our products compared to competitors' cost performance level?* These questions are important for manufacturing companies, both when (1) products are sold to end consumers, in the context in which competitors' cost levels are one of their best kept secrets, and (2) products are sold to non-end customers to know what their choices are among current or potential competitors.

Therefore, competitors' prices/costs are the ones that can decisively influence the target sales level and, implicitly, the target profit level throughout the life cycle of the products [from the initial research and development (R&D) stage to after-sales service and activities of product removal from the market]. Often managers know that if the cost level is not competitive throughout the product life cycle, then the company will leave the market sooner or later. In line with Ohno's previous formula, in order to answer the earlier questions, the starting point is the analysis of competitors' prices and, in particular, of the best competitor for each PFC and possibly for each product individually. The earlier formula can also be presented as such:

$$\text{Target sale price} - \text{Target profit} = \text{Target costs} \qquad (2.1)$$

Formula 2.1 is the basic formula for setting the target costs for new products throughout their life cycle (therefore for current products as well) (Kato, 1993; Cooper, 1996; Lee et al., 2002; Cooper and Slagmulder, 2004; Ansari et al., 2006).

From the pricing benchmark perspective, comparative companies are companies that have the best price in the market, or which have an immediate inferior price level for relatively similar quality products to those of the company carrying out the benchmark pricing survey. Figure 2.2 presents an identification analysis of the comparative companies for a "P" (or PFC) product (existing product or a future product). The product of the company performing the pricing benchmark is product "A"; products B, C, D, E, and F are the direct competitive products of comparative companies. The other products marked with "X" are not in the scope of analysis of comparative companies for the time being. Therefore, depending on the evolution of sales, in order to increase its profit level and competitiveness by price and cost, product "A" will seek to identify the competitive advantages by price/cost and quality of products "B" and "C" (if sales are rising) and of products "D" and "E" (if sales are declining). The product "F" is considered to be the product requiring constant identification of competitive advantages through price/cost and quality, regardless of the level of sales. Subsequently, from the perspective of the MCPD system, the competitive advantages of the manufacturing cost structure

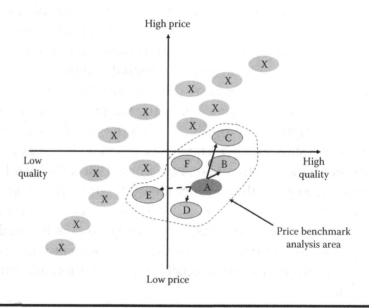

Figure 2.2 Consumer perception by price and quality and setting comparative companies for each product or family of products.

will be identified and systematic (kaizen) and systemic (kaikaku) improvement activities will be carried out in order to reduce the gap between the current state of the manufacturing cost structure and the one required (target).

Therefore, the price benchmark analysis seeks to establish a winning behavior of long-term price levels that will help to obtain a profit margin that will ensure a reasonable return on investment (ROI). Moreover, by knowing and influencing prices, we aim to achieve the multiannual profit targets, the external multiannual target profits and, consequently, the establishment of the internal multiannual target profits (by setting the annual MCI target in dynamics).

2.4.1.2 Price Benchmark Analysis for Each PFC

To meet the need for multiannual profit (target profit; see Formula 1.1) and annual profit (annual target profit; see Formula 1.2), manufacturing companies establish for each PFC and, possibly, for each product, throughout their life cycle, both for the new products and existing products, the following: (1) target profit, (2) target turnover, (3) target sales quantities, (4) target price, and (5) target costs.

Within this context, the benchmark of comparative companies for price structure and associated costs for the most important competitors is the starting point of the external analysis of the MCPD system as it seeks to identify the best strategic cost-improvement directions for each PFC and/or individual products, in order to continually target a harmonious transformation of the company through target costs to meet the target profit.

Figure 2.3 shows the *price benchmark analysis steps for each PFC*. The purpose of this analysis is to establish MCPD system (step 1) analysis directions originating from the company's external environment in order to develop robust price/cost competitiveness with comparative companies and, implicitly, reach the target profit.

In this context, the price and cost involved are critical success factors for the future transformation of the manufacturing company with the purpose of achieving target profit in an MCPD environment. Price benchmark analysis for each PFC is achieved in four steps, as follows:

1. *Product quantity analysis*: To direct efforts to identify price/cost competitiveness discrepancies of product models, we analyze the volumes of all products manufactured and sold in the past (because the level of stocks of finished products is very low). At the same time, a

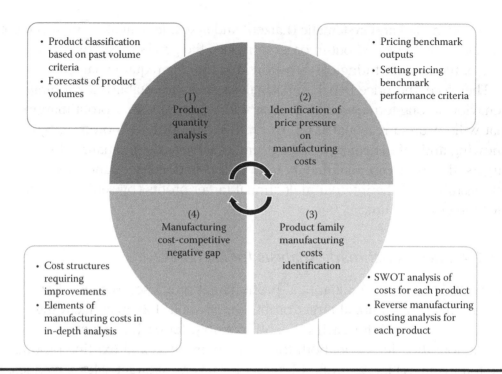

Figure 2.3 Price benchmark analysis steps for each PFC.

forecast with the highest accuracy of future sales volumes (possibly only those already contracted) is performed: the identification of products with the highest volumes. Figure 2.4 shows an example of product quantity analysis. As one can see, the total volume of the first 13 product models represents approximately 80% of the total production achieved and sold in year "N."

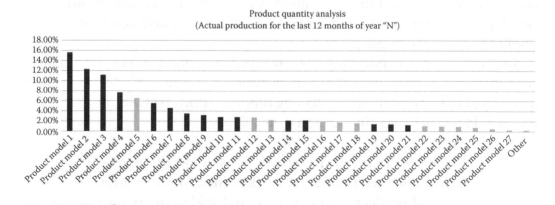

Figure 2.4 Product quantity analysis.

2. *Identification of price pressure on manufacturing costs*: For
   each product model of comparative companies, companies that
   have products approximately with the same characteristics as those
   of the company performing the pricing benchmark, we identify the
   following *price/quality performance measurement criteria* (but not
   limited to):
   - All current and/or potential brands
   - Basic and additional technologies
   - Current and future production capacities
   - Exploitation levels of available or future capacities
   - Design of their products and basic and additional functions
   - The quality of the raw materials and/or the components
     incorporated in the products
   - The main possible suppliers for each type of raw material and/or
     component and their pricing policies
   - Geographical locations of production
   - The probable cost level for each location of comparative companies:
     labor, utilities, taxes, cost of capital, cost of insurance, etc.
   - The geographic locations of potential suppliers of raw materials and
     components and the cost of transport
   - Sales volumes for each geographic area
   - Average selling price throughout the life cycle
   - Average selling price for each basic and additional part of the
     product reviewed

   These data are collected using different data collection and processing
   methods according to the specificity of each product model of
   comparative companies. Following this analysis, the following *pricing
   benchmark outputs* are obtained:
   - All product groups that are in direct and indirect competition.
   - The current and potential volumes of each comparative company
     market niche.
   - Basic and additional functions of comparative companies (including
     the most frequently and rarely offered functions).
   - The approximate weight of each function of the products in the
     sales prices practiced by the comparative companies for each
     product model.
   - The approximate cost of each function individually from the price
     structure (in particular the cost of raw material and/or components
     based on suppliers' pricing—especially for components).

■ Approximate price and cost differences for (1) each product model and (2) each basic function and/or additional function. This way, it is possible to identify the possible price pressures applied by the comparative companies on the cost structure of the functions of each product model (existing and/or future) of the company performing the pricing benchmark, including the life cycle trends of the products. Figure 2.5 shows an example of price benchmarking for a product model ("X") from company "A." As one can notice, the structure of the selling price of each competitor (comparative companies) is analyzed from the perspective of total costs. Determining the manufacturing costs associated percentage is the main purpose. This percentage is influenced by all other costs and by the percentage of costs unknown at the time of analysis (e.g., 2% for competitor 6). In this example, competitor 6 is considered to be in the *best practice cost (BPC)* situation because it practices a lower price than the product of company "A," almost all costs are known, the profit level of 10% is acceptable, and the life cycle of the two products

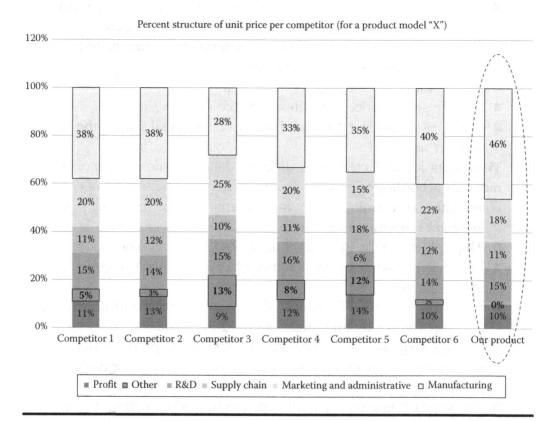

Figure 2.5   Price benchmarking for each product model: profit and cost structure in unit price per competitor vs. our product.

is almost similar. This way, they will seek to fully understand and assimilate all the means of constructing the costs used by competitor 6. From the perspective of manufacturing costs, we will continually seek to understand fully: the cost level of each raw material and components for two to three new possible suppliers, supplier lead time for two to three new possible suppliers and associated costs, technology used, product features and possible process parameters, possible management methods (other than current company "A"), the necessary performance of the production processes, the necessary skills of the employees, etc.

3. *Product family manufacturing costs (PFMC) identification*: Considering the product quantity analysis and the pressures on the manufacturing cost structure of each existing and/or future product model, the product family production cost is determined. Product models that are part of the same product family manufacturing cost are those that have about the same pressures on cost structures from comparative companies, have about the same types of manufacturing costs, and run approximately the same manufacturing flow processes to perform functions comparable to those of competitive product models. Determination of PFC occurs in two large concomitant stages:

   a. *Strengths, Weaknesses, Opportunities, and Threats (SWOT) analysis of costs for each product*: It is performed for each product model of the company that performs a pricing benchmark or product family as compared to the product models of each comparative company in order to identify the factors that can influence the dynamic evolution of future costs for existing or future products. An example of SWOT analysis of costs for a product family/product is shown in Figure 2.6. Further on, from a manufacturing cost perspective, a SWOT analysis of costs is performed on the entire manufacturing flow, especially by determining opportunities and threats in terms of costs for manufacturing costs by querying the costs of all the processes of each product family. For this purpose, graphs of time evolution of each cost structure are created and the trends necessary for achieving them are determined.

   b. *Reverse Manufacturing Costing Analysis (RMCA) for each product family/product*: It is performed to identify the product family manufacturing cost and the main types of costs that require further improvements to ensure the level of price/cost competitiveness and long-term profit level. *RMCA is determined as follows*:

      ■ *Product quantity analysis for each product family*: Product volumes that have approximately the same manufacturing flow

Strengths in terms of costs	Weaknesses in terms of costs
(*Examples*: • Low cost of electricity • Low cost of transportation of raw materials and components from suppliers • Low cost of transportation to customers • Low cost of packaging • Low cost with rework due to the transportation of finished products • Relatively low cost with capital, etc.)	(*Examples*: • The trend of increasing raw material prices • The high cost of scrap resulting from the transport of raw materials and components • The trend of increasing costs of new product development and design • The trend of increasing costs of intermediate storage of finished products • The trend of increasing costs with bank financing, etc.)

SWOT analysis
of costs for
each product

Opportunities in terms of costs	Threats in terms of costs
(*Examples*: • Improving the cost of losses for: (1) equipment, (2) human work, and (3) materials and utilities • Improving the cost of work in process (WIP) • Improving the direct labor costs (standard labor time) • Improving maintenance costs (standard maintenance time) • Improving general administrative costs)	(*Examples*: • The trend of increasing costs of energy losses • The trend of increasing costs of breakdown losses • The trend of increasing costs of equipment rework losses • The trend of increasing costs of setup losses • The trend of increasing costs of spare parts and lubricants)

Figure 2.6   Example of SWOT analysis of costs for each product family/product.

processes are determined. Figure 2.4 shows two families of products: product family 1 with 78.02% of the total quantities in the past (product models 1, 2, 3, 4, 6, 7, 8, 9, 10, 11, 14, 15, 19, 20, and 21) and product family 2 with 21.98% (product models 5, 12, 13, 16, 17, 18, 22, 23, 24, 25, 26, and 27).

■ *Price pressure exerted by comparative companies*: The cost types that are at a disadvantage compared to products belonging to comparative companies are identified. From Figure 2.5, it can be seen that our product has a manufacturing cost of 46% of the price structure and that it can best compare to the price structure of competitors 2 and 6. These competitors have a high accuracy of price structure and an acceptable level of profit.

4. *Manufacturing cost competitiveness negative gap identification*: Based on PFMC, the differences in *cost structures that require improvements* to determine a competitive price and to ensure an acceptable level of profit are determined. In the example in Figure 2.5, it can be seen that the manufacturing costs of 46% pose a problem. For this, the comparative

analysis is performed in detail for *cost of the goods sold* (cost of raw materials and components, direct labor costs, manufacturing overhead). For the other cost structures, the comparative analysis is performed for *operating expenses* (R&D costs, costs of designing and redesigning manufacturing processes, marketing costs, distribution costs, post-sale services costs, etc.). Based on *SWOT analysis of costs*, the costs that will be under analysis in the next period, their location in processes/ physical areas, and the differences from those of comparative companies (*manufacturing costs in in-depth analysis*) are established.

2.4.1.3 Current and Future Competitive Gap for Each PFC

In order to determine the current and future manufacturing competitive gap, it starts from the necessary level of price reduction for existing and/or future products based on marketing analyses, a reduction that would ensure acceptable competitiveness and the achievement of the long-term profit target and of the annual targets profit.

To exemplify how to establish a current and future manufacturing competitive gap analysis, the earlier example of company "A" (see Figure 2.5) shall be continued. Therefore, company "A" (the one carrying out the analysis) expects the price level to decrease in the following period by 7% for *product family 1* (*for the next year* "N + 1"). The top management from company "A" wants to ensure acceptable competition and plans to reduce the price in average by 10% for all *product family 1* products, from an average of $100 to $90 per unit. If this price reduction were to be achieved, the marketing manager from company "A" predicts sales growth from 700,000 to 925,000 units and, implicitly, a profit of approximately $92,500 (10%). In this context, the total annual target revenue is $83,259,000 (90%*925,000 pieces), the total target profit (10%) is $8,325,000, the unit target profit is $9 ($8,325,000/925,000 pieces), and the target cost per unit is $81 ($90 − $9). The current total costs for product family 1 are $66,500,000, and the current unit cost is $95 ($66,500,000/700,000 pieces). Further on, the *projected future manufacturing cost performance levels* are established based on the current and future manufacturing competitiveness negative gap analysis for each product family. Table 2.1 shows the level of the annual MCI goal (based on MCI targets and means). Concurrently or not, in order to achieve total cost reduction management, cost reduction targets and means of achieving them are set for R&D (*R&D cost policy deployment, R&D CPD*), for the supply chain (*supply chain cost policy deployment, SCCPD*), and for the marketing

Table 2.1 Full Unit Target Profitability

	Current Price		Current Costs		Target Price		Target Costs	
	%	Value ($)	%	Value ($)	%	Value ($)	%	Value ($)
R&D and design costs	15	15	15	14.3	13.7	12	13.7	**11.1**
Supply chain costs	11	11	11	10.5	12.5	11	12.3	10.0
Marketing and administrative costs	18	18	18	17.1	13.2	12	13.2	10.7
Manufacturing costs	46	46	46	43.7	50.6	46	50.7	41.1
Target profit	10	10	10	9.5	10.0	9	10.0	8.1
Total	100	**100**	100	**95**	100	**90**	100.0	**81**

and administration (*marketing and administrative cost policy deployment, MACPD*). As one can observe, for the year "N + 1," the MCI goal is about 6% (a reduction from $43.7 to $41.1). In order to achieve the $81 target cost, R&D costs and marketing costs will be reduced, because in the year "N" a number of new important products have already been developed and launched.

To achieve the *total cost policy deployment* (*TCPD*), there is a need to establish targets and means for costs (cost policy) across the entire supply chain beyond the processes of a manufacturing company for all business stakeholders who are involved in the final price formation to identify all opportunities for cost reduction and including consistent and sustainable price.

2.4.1.4 Knowing Competitive Ways for Each PFC

The fulfillment of the long-term target profit for each PFC, especially the manufacturing profit (external and internal), requires the coordination of the manufacturing costs according to the *signals transmitted continuously by the market* (the number of finished products needed to be manufactured and sold—both existing and new, the dynamic evolution of average prices, the required stock level of finished products, units sold daily, etc.) and the *need for acceptable development of long-term manufacturing companies* (acceptable unit profit, production/manufacturing capacity, the number of new products released annually, number of new equipment/technologies implemented annually, etc.). In this context, the main role of manufacturing costs is to continuously sustain and stimulate sales, based on a market-accepted cost, with other cost structures outside the manufacturing area throughout the life cycle of every product of each PFC, on the basis of ensuring a level of unit cost improvement, with consistent productivity and quality throughout the entire manufacturing flow.

To meet the market needs, the continuous improvement of manufacturing costs is indissolubly linked to the continuous improvement in productivity and quality level. In order to continuously know the possible destination of the manufacturing cost, starting from the market price level and the required long-term profit level, it is necessary to continually anchor the manufacturing cost information in the logic of performance management related to productivity and quality in particular. More precisely, the manufacturing cost policy (targets and means) deployment for each PFC and for the entire company is consistently connected with all the manufacturing company policy deployment reasoning. Therefore, the need to *meet MCI targets or manufacturing cost decrease* (material costs reduction and transformation costs reduction) must be continuously combined with the following:

■ *The need to reduce all other costs throughout the life cycle of products and/or equipment* (design costs reduction, supply chain costs reduction, marketing and administrative costs reduction, reducing post-sale costs, reducing the cost of eliminating products/technologies)

■ *The need to continuously increase the company's speed and agility* (reducing the level of all inventories, reducing the lead time, increasing the efficiency of each area, reducing the variations between the standard and current time cycles, reducing small stops, increasing the use of hybrid components for more products, increasing the number of new products and reducing their time-to-market)

■ *The need to increase the number of delivered products* (productivity increase, product range increase, capacity increase)

■ *Increasing customer-required quality level* (increasing customer satisfaction and increasing product quality ratio)

From the perspective of the MCPD system, knowing and reaching on a continuous basis competitiveness manufacturing costs destination for each PFC is possible to be ensured through a robust system of productivity and quality throughout the entire manufacturing flow. Figure 2.7 presents the main objectives of the annual MCI means to meet the annual MCI targets for each PFC and, especially, the annual MCI goal: *product number*

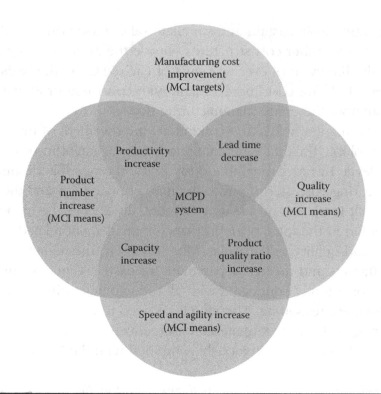

Figure 2.7 Road map for knowing and reaching the perennial destination for MCI of each PFC.

increase, speed and agility increase, quality increase. As one can see, these destinations are perennial; they do not have a final point in the journey.

Therefore, in order to meet the long-term manufacturing profit plan and an acceptable level of price throughout the life cycle of products and/or product families, establishing and achieving the required productivity level, the necessary capacity, lead time, and product quality ratio along the entire manufacturing flow represents the destination for each PFC from the MCPD system perspective. By setting the annual MCI targets continuously linked to market needs (price and profit) and by the four destinations (annual MCI means), a harmonious transformation of manufacturing companies can be achieved in order to ensure annual MCI goal and high competitiveness.

2.4.2 Synchronization of Life Cycle Targets for Each PFC

From the perspective of the MCPD system, companies continuously set and analyze the target levels related to revenues, profit, capacity, productivity, and cost over the entire life cycle of products (over many years). In order to establish an acceptable multiannual and annual level of target manufacturing costs, it is necessary to know the potential contribution to profit of each product and/or each PFC, regardless of the stage of each product. The interdependent synchronization of these targets is necessary to achieve the multiannual and, especially, annual MCI policy deployment plan.

To establish an annual MCI target with the most precise accuracy, intercorrelated multiannual and annual synchronization is required at the level of each process of each PFC, especially the following elements:

- *Target sales volumes* for each market individually and overall
- *The dynamics of the target price level* necessary for the sale of products for each market
- *The dynamics of the number of target products* needed to be sold to achieve the *targets revenues*
- *Planning the multiannual manufacturing target level* required for each year and each month in particular to support sales volumes
- *Planning the level of target productivity* needed for equipment, people, raw materials and materials, and utilities to support the dynamics of the target price level through the cost level and, implicitly, the profit plan

The lack of ownership of this continuous synchronization between the company exterior (target sales volumes, target sales prices, number of target products) and the company interior (target manufacturing capacity and target productivity) creates the premise of a high level of costs, both in the manufacturing stage and especially in the stages of marketing, distribution, supply, and after-sales services. From the manufacturing perspective, the lack of this synchronization often leads to an unstable manufacturing lead time, to constant overloading to achieve the cycle time synchronization from processes to takt time, to the instability of the standard level of WIP together with the instability of the standard level of consumables (spare parts, auxiliary materials, lubricants, etc.), and to the impossibility of fulfilling the continuous flow and one-piece flow principles. In this context, the level of productivity tends to decrease and the manufacturing cost and other costs to increase.

Planning the level of capacity needed to deal with sales volumes is achieved by planning the required productivity level and, implicitly, by planning to achieve MCI means (especially for: systematic improvement—kaizen and systemic improvement—kaikaku). Therefore, the symbiosis between kaizen and kaikaku aims to provide a capacity to deal with likely sales volumes by ensuring productivity that would enable an acceptable level of costs. In particular, by continually setting annual targets for kaizen and kaikaku projects (annual MCI means targets), the continuous synchronization with annual MCI targets is ensured at each PFC level and, implicitly, an acceptable level of company profit (both at the annual external profit and especially at the annual MCI goal level).

2.4.3 Efficiency of Investment and Productivity

The teams of managers often face the challenge of balancing long-term and short-term decisions based on financial, but especially nonfinancial, analyses. One such decision is to provide an efficient capacity to ensure the number of products required by the market. From the perspective of the MCPD, where the required capacity is mainly provided by the equipment, reasoning related to the capacity needs is strongly related to the increase in the current capacity level of the equipment (by kaizen) and the required capacity level of the future equipment (by kaikaku).

In the short term, before considering the feasibility of replacing any existing equipment with a new one, *amid a growing demand from the market*, a maximum allowable level of productivity should be achieved by

- Reducing and stabilizing the cycle time (in seconds or minutes)
- Increasing the Overall Equipment Effectiveness (OEE) level (as a percentage)
- Reducing the setup time (in minutes)
- Reducing the man*hours/product
- Reducing scrap rates and rework (as a percentage)

Also in the short term, *faced with a downward demand from the market*, it is necessary to maximize productivity, especially reducing the cost of raw materials and staff.

All these concerns of increasing the productivity in the short term are made with the help of feasible projects of systematic improvement (kaizen) or, more precisely, with the help of the productivity obtained within the company (*internal productivity*).

In the long run, replacement of equipment becomes necessary to ensure a superior level of capacity and productivity (including a better manufacturing cost). Irrespective of the method used to calculate *ROI* [*return on investment = (gain from investment − cost of investment)/cost of investment, return on investment = (revenue − cost of goods sold)/cost of goods sold*, or other forms of calculation] or its extensions, the need for productivity designed for the future equipment is necessary to ensure an acceptable feasibility of productivity gained from the outside of the company (*external productivity*). Therefore, setting targets for future equipment, besides the necessary productivity (cycle time, OEE, setup time, man*hour/product, scrap and rework rate, etc.), requires an initial cost analysis of the investment and a *time-to-start-production* (*TTSP*) analysis (in weeks or months). Any exceeding of the *TTSP*, or more precisely the time planned for the new equipment to be put into operation, draws a lost opportunity to ensure the required capacity and the target cost level planned to keep the profit plan (target profit).

In this context, the two paradigms of thinking, *return on investment and productivity*, must be viewed in a continuous symbiosis. Both aim to achieve the long-term profit plan by meeting the annual MCI goal.

2.5 Alignment of Improvement Opportunities to the Need for MCI

The primary purpose of any manufacturing company, and beyond, is to provide products and services at a price that is perceived as acceptable for a customer segment or for a particular market. Providing an acceptable price requires an acceptable unit manufacturing cost along with other cost structures (sales and administrative costs and all other costs along the entire supply chain to the final consumer) through the efficient and effective use of all current capacities through the continuous increase of daily production and, implicitly, by increasing the productivity of equipment, people, materials, and utilities and by promoting the activities of systematic improvement (kaizen) among all employees of the company and beyond.

Continuous identification of opportunities to improve manufacturing costs requires continuous measurement of the productivity and cost levels throughout the manufacturing flow and, implicitly, of all processes and activities with the purpose of identifying the optimal time to trigger kaizen activities to

- Reduce and/or permanently eliminate all non-value-added activities and all unnecessary consumption over the entire manufacturing lead time (found in losses and waste)
- Conduct studies and scenarios to reduce manufacturing unit cost by increasing the adding value to each activity
- Reduce and/or eliminate all non-value-added costs by actually reducing manufacturing lead time

For the annual and multiannual improvement of manufacturing unit costs and, implicitly, the achievement of an acceptable annual and multiannual profit (annual and multiannual target profit), a cost-all approach is needed, especially for transformation costs and material costs but also for design costs and depreciation costs, through the participation of all departments of the company and beyond. Knowing and stabilizing the manufacturing unit costs for further continuous improvement requires stability through standardization of equipment, people, inventory, quality assurance, and office activities. This stability, standardization, and continuous improvement is achieved by continuously targeting kaizen projects to these goals (Standard Operating Procedure—SOP). For example, for an annual MCI goal of 1% for a PFC, a target of a 2% reduction in transformation costs and a 3% reduction in direct material costs for all processes can be set. Furthermore, in order to achieve the 2% reduction target of the transformation costs, a target to increase OEE by 2% of the PFC process could be set by planning and rerouting two distinctive kaizen projects (reduce breakdown times by 1% and eliminating the differences between current and standard time cycles—speed down losses). In order to achieve the 3% direct material costs reduction target, targets can be planned to reduce unnecessary material consumption by 1% in all raw material transformation processes in finished products and to identify the price reduction opportunities from suppliers with 2%, possibly changing suppliers. This *manufacturing targets cost deployment* determines the number and percentage of employees' participation in achieving the unit manufacturing target costs. The same reasoning in the earlier example is used for multiannual profit planning. The major challenge is to know, on a continuous basis, the data on price levels and potential market volumes and to continually measure the productivity level (losses and waste) in order to know the most feasible opportunities for improvement and to develop, in a harmonious way, scenarios for meeting the annual MCI targets.

Consequently, from the perspective of the MCPD system, all opportunities for productivity improvement are planned and realized in such a way as to ensure the annual MCI goal (through cost savings/avoidance) and the annual external profit (by increasing the capacity to support sales required by the market). These annual judgments converge with multiannual profit target and, implicitly, with productivity vision.

References

Ansari, S., Bell, J., and Okano, H., 2006. Target costing: Uncharted research territory. In Chapman, C. S., Hopwood, A. G., and Shields, M. D. (Eds.), *Handbooks of Management Accounting Research*, Vol. 2. Amsterdam, the Netherlands: Elsevier, pp. 507–530.

Cooper, R., 1996. Costing techniques to support corporate strategy: Evidence from Japan. *Management Accounting Research*, 7(2), 219–246.

Cooper, R. and Slagmulder, R., 2004. Interorganizational cost management and relational context. *Accounting, Organizations and Society*, 29(1), 1–26.

Kato, Y., 1993. Target costing support systems: Lessons from leading Japanese companies. *Management Accounting Research*, 4(1), 33–47.

Lee, J.-M., Chen, I., Chen, R. C., and Chung, C. H., 2002. A target-costing based strategic decision support system. *Journal of Computer Information Systems*, 43(1), 110–116.

Ohno, T., 1988. *Toyota Production System: Beyond Large-Scale Production*. Cambridge, MA: Productivity Press.

Posteucă, A., 2011. Management branding (MB): Performance improvement through contextual managerial behavior development. *International Journal of Productivity and Performance Management*, 60(5), 529–543.

Posteucă, A. and Sakamoto, S., 2017. *Manufacturing Cost Policy Deployment (MCPD) and Methods Design Concept (MDC): The Path to Competitiveness*. New York: Taylor & Francis.

MCPD TRANSFORMATION

Chapter 3

Establishment of an MCPD System: Steps to Begin

The fundamental tasks of any manager are to continually understand the real situation of the company's external and internal business environment, to identify the main issues alone, and to come up with solutions and to mobilize his/her own staff, colleagues, superiors, and all other persons from outside the company to achieve the expected results. Managing cost reduction means understanding, planning, and gaining control over all the activities needed to have efficient and effective processes, to achieve your goals and meet a cost targets.

Before deciding on the adoption of the manufacturing cost policy deployment (MCPD) system for the achievement of the long-term target profit and the annual target profit, the initial steps concern the preparation of the management system for the establishment and fulfillment of the annual manufacturing cost improvement (MCI) goal and, implicitly, the annual MCI targets and means, and for strategic and departmental organization for achieving the annual MCI goal to complete the annual target profit, besides the annual external profit (from sales).

Firstly, it is necessary to measure in detail the essential dimensions of the whole company, both its *external*, market (market driven for MCI) and profit (profit driven for MCI), *dimensions* and *internal dimensions*, the manufacturing costs (driven by MCI targets deployment) and the manufacturing flow (driven by MCI means deployment). These measurements start from the vision and mission of the company on productivity, implicitly from the current and expected market share, from the long-term profit plan of the overall company (profit target) and the profit plans (profit target) of each product family cost (PFC) throughout their life cycle.

In order to motivate people to adopt the MCPD system, they need to understand from the beginning the main implications of changing the current paradigm to address the continuous improvement of the manufacturing flow by focusing on MCI. The managers need to ask objective questions about the manner of occurrence and manifestation of CLW and CCLW and not hide the real size of the profit margin from the losses and waste categories of each process and, especially, the causality relationship between losses and waste along the manufacturing flow.

The core stakes of managers to adopt the MCPD system lie that by achieving the annual MCI goal, the following are achieved concomitantly:

1. Multiannual target profit consistently (external, but especially internal profit)
2. Continuous reduction of the unit manufacturing cost level, which ensures the continuous development of the company by increasing the production volumes and, implicitly, the target market share
3. Continuous improvement of operational performance throughout the entire manufacturing flow

All these stakes, achieved through the MCPD system, contribute to practicing a performance management that ensures a sustainable transformation of manufacturing companies. At the same time, people's attitudes and behaviors are changed at all levels against the improvement opportunities as their initiatives are continually aligned with the strategic needs for achieving the annual MCI goal; they are directly coordinated by the real market needs of manufacturing costs reduction and, implicitly, ensuring on time all the necessary resources.

Only after the acceptable preparation of the implementation of the MCPD system, about three to six months, depending on the size of the company, one can proceed to the detailed development of the three phases and the seven steps. The essential elements of the initial preparation of the MCPD implementation are (1) the full involvement of the management team and (2) establishing the connections between the desired medium- and long-term effects (what do we want?) and the MCPD system processes (how to approach through MCPD?).

So, this chapter gives answers to the following question: *what are the steps preceding the introduction of the MCPD system?* The answers are two ways: a technical answer (MCI targets deployment: preliminary steps) and an organizational answer (preparation for the implementation of an MCPD system). Therefore, in Section 3.1, the preliminary steps needed to establish MCI targets are presented by approaching in detail the market-driven activities for setting annual MCI goal, the profit-driven activities for setting annual MCI targets, and the annual management coordinated by MCI targets and means deployment. Then, in Section 3.2, the organizational steps necessary for the adoption of the MCPD system are presented, as a way of fulfilling the multiannual production profit plan by productivity improvement; more precisely, the elements underlying the MCPD system implementation decision and the prediction of its effects, the MCPD system organization, and the creation of its structures to support MCI and internal and external communication of the MCPD system purpose are presented.

3.1 MCI Targets Deployment: Preliminary Steps

At the level of a group of companies or a company within a group, the annual challenge of the MCPD system is setting the annual MCI goal to meet the annual target profit (see Formula 1.2). The setting of the annual target MCI should consider the multiannual targets profit (implicitly the level of external and internal profit—see Formula 1.1) and the real possibility of the contribution of each PFC to this goal, through MCI targets and means at the level of their processes.

MCI targets deployment at PFC level aims at ensuring concomitant customer satisfaction by price/cost, quality, and delivery times. In order to accomplish all connections and stratifications at Overall Management Indicators (OMIs) and Key Performance Indicator (KPI) levels (10 steps ahead of MCI targets and means; see Section 1.4), the management teams are and will continue to be continuously concerned of ensuring a high level of synchronization for all process capacities at the pace of customer demand (takt time) by providing the highest possible OTIF (On-Time In-Full) for the delivered finished products, as close as possible to 100%. In order to achieve the highest possible percentage of OTIF, companies develop strategies and action plans for the optimization of (1) *material lead time, MaLT*; (2) *factory lead time, FLT* [*manufacturing lead time (MLT)* and *stock days for raw material and finished products*]; and (3) *delivery lead time, DLT.* From the perspective of MCI targets deployment, the substantiation of these strategies and action plans is based on a detailed analysis on the current state of the overall historic development of the company and, for each PFC, on (1) the need to reduce the unit manufacturing costs to meet the annual MCI goal; (2) the need to increase productivity; (3) the total level of losses and waste; (4) the total level of losses and waste for Manufacturing Key Points (MKP); (5) the total level of CLW for each important process/work center/activity or, for all; (6) the overall level of CCLW for the Manufacturing Cost Key Points (MCKP) (the root cause of CLW and, implicitly, of associated losses and waste, upstream and downstream of the entire manufacturing flow and beyond); (7) the effectiveness and efficiency of kaizen and kaikaku projects completed to achieve the annual MCI targets and further on the annual MCI goal; and (8) the current performance of kaizen and kaikaku projects underway.

The FLT optimization requires the continuous reduction and/or elimination of losses and waste. The main MLT concerns are related to (1) planning and ensuring the capacity to ensure the availability, speed, and acceptable quality of the manufacturing processes from the perspective of the *maximum level of takt time* and (2) a level of work in progress (WIP) approaching as much as possible the *one-piece flow* (OPF) state. The fulfillment of these concerns creates the prerequisites for the fulfillment of the *zero costs of losses and waste* (ZCLW) state.

From the perspective of meeting the annual MCI goal for each PFC in part and on the total plant, the major concern is to continuously identify, reduce, and/or eliminate all associated costs and waste from the MKP (takt time maxim/bottleneck cycle time) and processes that scatter most CLWs across the entire production flow (*CCLW—the acute cost problems*). By continually targeting all systematic and systemic improvement efforts to reduce and/or eliminate these CCLWs, implicitly their effects on all other manufacturing costs and beyond for each PFC, and by addressing the significant CLW from certain processes (*the chronic cost problems*), it is possible to eliminate all obstacles to achieving the annual MCI goal through annual MCI targets and, implicitly, of the annual and multiannual target profit.

Therefore, starting from market signals and the need for the annual manufacturing target profit and the annual MCI goal, MCI targets deployment will aim the setting of targets for the following:

■ CCLW, which implies time, *time-related losses—TRL* (costs of equipment losses and costs of human work losses)
■ Certain significant CLW, which implies *physical losses, PL* (with high level in a particular process/work center, but which do not extend too

clearly throughout the entire manufacturing flow), for example, costs of technological and material scrap losses; cost of auxiliary consumables losses; costs of energy losses; costs of die, jig, and tool losses; costs of finished products stock, waste (exceeding the number of stock days standard for finished products); costs of raw material stock, waste (exceeding the number of stock days standard for raw material); and costs of components stock, waste (exceeding the number of stock days standard for components) (Posteucă and Sakamoto, 2017, pp. 116–122).

Next, we will outline the preliminary steps of the annual MCI targets deployment for a PFC. It would be desirable that the annual MCI means deployment be achieved in parallel with the annual MCI targets deployment or as sooner as possible after establishing the MCI targets deployment to achieve the annual MCI goal of a PFC.

3.1.1 Market Driven for Annual MCI Goal

Continuous alignment of the annual MCI goal to market pressures and requirements from a price target perspective plays a critical role in setting the annual MCI targets at the level of PFC processes. Price benchmark analysis and synchronization of life cycle targets for profit, price, sales, and necessary capacity are the ingredients that continuously capture the market pressure over annual MCI targets and, furthermore, on the annual MCI means. This pressure is taken over by the entire company, not just by the *FLT.*

Market driven to set the annual level of MCI goal involves the detailed analysis of the following items:

■ *Long-term sales level and company target market share (productivity vision and mission)*: The purpose of this analysis is to ensure that each existing product and each future product will, throughout its lifetime, contribute to the company's long-term sales and profits plan. The analysis of competitors and customers, conducted by the marketing department, is carried out for the evolution of sales, average prices, and market share for the last representative period (usually the last three years or starting with the moment when this monitoring was started) for each product and/or for each PFC, and for projections for the next three to five years. The stake of these analyses is to predict with the highest accuracy possible the sales volumes and, implicitly, the required annual production capacities (production numbers). The realistic analysis of the behaviors and trends of customers and competitors will constitute the basis for establishing the production capacities necessary in the future and the annual MCI targets and, implicitly, the basis of establishing the need for improving the current capacities (kaizen projects—MCI mean level 2) and increasing current capacity through investments (kaikaku projects—MCI means level 3). In this way, the annual MCI targets and means directly contribute to increasing the annual external profitability level and long-term consistent competitiveness, even if the intensity of the competition is high and customer behavior is changing and sophisticated.

■ *Product structuring within each PFC*: To satisfy each and every client, in fact the widest possible pool of customers, in terms of price and, implicitly, costs (in the case of MCPD through unit manufacturing costs), it is necessary that the marketing department identifies the impact of customer and competitor behavior over costs. These behaviors relate especially to (1) the rhythmicity and magnitude of the price reductions applied to similar or relatively similar products for each customer segment and (2) the level of perception of the costs associated with each basic and/or additional function. As mentioned earlier, PFC means families of products that have the same market pressure on the need for annual MCI goal; they run around the same manufacturing flow processes and have roughly the same opportunities for the annual MCI means and the same target for MCI. The purpose of this analysis is to understand any type of potential customer and competitor behavior for the next 12 months, and especially for the next six months, which may impact on the annual external profit reduction that will lead to an increase in pressure over the annual MCI goal during an already budgeted year in which the annual target profit level has already been set. Annual target profit is the promise of the management team at the beginning of the year to be complied with at the end of the year.

■ *Capture price evolution and set target selling price (average per seasons and product life cycle stages)*: The marketing department analyzes the sales potential of current and future products by practicing the market prices and securing the multiannual target profit level for each PFC and for the entire company. The analysis of the evolution of the past average price for each product and/or for each PFC is performed considering (1) the value perceived by customers for each basic and additional function of each product and (2) the contribution to the profit brought by every product existing on its life cycle. Any variation in customer-perceived value with impact on sales volumes and the level of each product's contribution to the company's long-term profit plan (multiannual targets profit), especially the negative variations, results in a realignment of pressure on the annual MCI goal and, further on, the annual MCI targets and means for the processes of each product and/or the processes of each PFC. Furthermore, the pressure on the annual MCI goal is further accentuated by the future level of target price suggested by the marketing department. This target price level for sale, usually lower than the current one, is supposed to help increase sales volumes. Therefore, by continuously capturing the evolution of past average prices, considering the seasonality and product life cycle, the current status of the contribution of each product and/or PFC to the multiannual profit targets is determined, which together with the future price level for the future suggested by the marketing department will determine a realistic level of pressure on the annual MCI goal for each product and/or for each PFC. Generally, any attempt to raise sales prices is toned down by the prompt reactions of competitors and, above all, customers. In this context, it is necessary to reconcile, through successive simulations, the evolution of prices and sales volumes in the past and their forecasts and the level of multiannual and annual target profit, especially with the level of the annual MCI goal achieved by the annual MCI targets and means at each PFC level by exploiting the hidden reserves of profitability (MCI targets level 1 reconciliation).

3.1.2 Profit Driven for Annual MCI Goal

The continued alignment of the annual MCI goal to the need to achieve a multiannual acceptable profit (target profit) to ensure the sustainable development of the manufacturing companies is again a critical point of setting the annual MCI targets. Even if the annual profit percentage of a group of companies is relatively stable (for three to five years), for example, 10% profit margin (*profit margin = net income/net sales or revenue*), this percentage needs to rely on the exact sources at the level of each year, at the level of each company in the group, at the level of each PFC, and at the level of each product throughout their life cycle, through *successive simulations* of the annual profit from sales (annual external profit), by meeting expected sales volumes at the market price and profit from improvements (annual MCI target), through the achievement of the annual MCI targets and means.

Fulfilling the MCI targets and means will ensure both sales volume fulfillment, by continually improving the required capacity level at the right time, and meeting the required unit manufacturing cost reductions. This is the main stake of the MCPD system: supporting both the annual external profit and especially the annual MCI goal. Therefore, the continued transformation of manufacturing companies coordinated by the need of cost reduction for uncovering hidden reserves of profitability will aim to achieve the annual external profit by increasing capacity, especially those available, and meeting the annual MCI goal by continuously reducing the unit manufacturing costs and, implicitly, by meeting the annual MCI targets and means (for each PFC and overall company).

Since the MCPD system is not a costing system, it depends on the accuracy of cost-related data and information provided by the cost and managerial accounting. The clear definition of input data on manufacturing costs is required to have an acceptable accuracy in terms of setting the annual MCI goal and, especially, the annual MCI targets at each PFC process (at each cost center) level.

3.1.2.1 Long-Term and Annual Realistic Manufacturing Profit Expectations

Starting from the main ingredients of the long-term manufacturing target profit presented in Section 2.4 and the need for alignment of the improvement opportunities for MCI presented in Section 2.5, realistic long-term manufacturing profit targets are set at the group level, at the level of individual companies within the group, and at the level of each PFC or for each product.

The strategic approach to establishing long-term realistic manufacturing profit expectations takes place as such:

■ The multiannual target profit is established, for example, $100 million for the next five years.
■ The expected level of external profit is set, for example, $80 million for the next five years.
■ The difference is established, in our example, $20 million.
■ We seek to cover the difference either by increasing the expected external profit level or by setting the internal profit level (by setting an annual MCI goal). If the sales level has a steadily rising trend, then 50% of the $20 million difference ($10 million) may be supported by the increase in external profit (from sales) and 50% of the difference to be supported by the internal profit (through annual MCI goal). If the level of sales is decreasing, then the share supported by the internal profit (through the annual MCI goal) increases, for example, to 60%–65% of

the $20 million difference (the emphasis on the annual MCI targets is on reducing material costs and reducing direct and indirect labor costs). If the level of sales is increasing, then the share supported by the external profit (by increasing the sales volume and, implicitly, the current and/or new capacities) increases, for example, to 60%–65% of the $20 million difference (the emphasis on the annual MCI targets being placed on reducing transformation costs and, implicitly, on increasing overall equipment effectiveness (OEE) and annual MCI means—kaikaku).

To this end, two types of successive simulations of external profit (from sales) and internal profit (achieving the annual MCI goal) are used.

The first simulation method starts from the current manufacturing profit of each component product of the PFC and continues with the target manufacturing profit for the next three to five years for different sales levels of current and future products. Sales volume status is taken into account: decreasing or increasing.

The elements of simulating the profit margin on sales (external profit), three to five years and annually, are future sales volumes of current and future products, the level of future prices of current and future products, and the level of manufacturing costs of current and future products.

The elements of simulating the internal profit in three to five years and on an annual basis are as follows:

■ *The level of future sales volumes* of current and future products
■ *The current average price level* of current and future products
■ *The target average price level* of current and future products
■ *The total manufacturing costs* of current and future products
■ *Total annual MCI goal achievable* for current and future products (based on total CLW identified at the level of each process/work center of each PFC)
■ *Total annual MCI targets* for current and future products (based on CLW and CCLW that are efficient and effective to be obtained through the annual MCI means)

For example, for calculating internal profit for one year for a single PFC at company level, if

■ The average sales volumes are planned for 3 million pieces/year.
■ The current average unit price is $5.5/piece.
■ The planned annual revenue level is $16.5 million/year.

- The standard cost is $5/piece (at the level of current standard costs).
- The level of annual planned total manufacturing cost is $15 million (at the annual budget level).
- The annual external profit expected is $1.5 million (3 million pieces/year * $0.5/piece).
- The unitary current profit is $0.5/piece—external unit profit.
- The target average unit price level is $4.8/piece.
- The annual profit target level is $2.1 million.
- The annual net MCI goal level is $600,000 (taking into account some risks of noncompliance of the annual MCI targets).
- The total maximum annual MCI goal to be achieved is $5.25 million (by identifying the total CLW and CCLW at the level of each PFC process/work center) or 35% of the planned annual total manufacturing costs of $15 million.

Then, it can be reached to set a total level of the annual MCI goal of $2.75 million (by choosing CLW and CCLW improvements at process level and by reconciling the annual MCI targets with the annual MCI means) to cope with the need for unit manufacturing cost reduction and, implicitly, the unit target price and to meet the annual target profit expected at $2.1 million. The total annual MCI goal of $2.75 million or 18.33% of the planned annual level of total manufacturing cost is considered acceptable and achievable to support the annual profit target together with an annual external profit of $0.5/piece.

If the annual MCI goal would be met, then the unit manufacturing cost would drop from $5/piece to $4.08 ($15 million − $2.75 million = $12.25 million; $12.25 million/$3 million pieces = $4.08/piece) that would lead to an annual unit manufacturing target profit of $0.72/piece ($4.8 − $4.08). The difference between the current unit profit of $0.5 and the target unit profit of $0.72, at $0.22/piece, can be considered acceptable as some of it may be considered to cover some of the risks that may arise from the failure to meet annual MCI targets, in our example of $0.02/piece, and the remaining $0.2/piece is the plus of profit earned from meeting the annual MCI goal and, implicitly, the annual MCI targets and means. Therefore, the annual net MCI goal is of $600,000 ($0.25/piece * 3 million pieces). In this context, the annual target profit is reached at the level of $2.1 million ($1.5 million of external profit/budgeted in the master budget and $600,000 of internal profit/budgeted in the annual manufacturing improvement budget). This annual target profit converges with the multiannual target profit. Meeting the annual MCI goal and, implicitly, decreasing the selling price will generate an increase in the annual external profit as a result of the increase in sales volumes amid a fall in unit price to $4.8/piece. In the following year, the total maximum annual MCI goal to be reached of $5.25 million will change as other internal realities (the level of losses and waste at the level of PFC process) and implicitly other levels of CCLW and of CLW will be captured. As a result, another level of the annual MCI goal will be set to meet the annual target profit. This approach is also used if there are several PFCs in the company. The contribution of each PFC to the annual target profit will be determined by summing the annual MCI goal of each PFC.

This is the annual stake for the MCPD system. It actually represents the increase of competitiveness through price and through uncovering hidden reserves of profitability.

The cost/benefit analyses of the annual MCI means (kaizen and kaikaku) will refer to the annual MCI targets to allocate all time resources needed to meet the annual MCI goal. When calculating the standard unit cost, the costs related to the annual improvements are also calculated (it is the budget structure related to the annual improvement budgets—supplemented by a master budget; this is another reason why MCI targets and means are set at the same time; therefore, the profitability simulations in the long run will also take into account the cost of improvements). The annual MCI targets deployment, which is grounded on annual MCI means deployment, is achieved through the exact location of the annual MCI target of $2.75 million at the process/work centers/activities level of the PFC. Simulation only stops when the annual MCI goal level and the target unit price levels are not met for the PFC. Simulations continue at the level of each PFC process/work center and at the level of each type of losses and waste until the most realistic annual MCI goal is identified based on the annual MCI targets and means. Annual MCI targets and means will aim at establishing the annual MCI goal participation to the annual target profit through annual improvements in transformation costs and material costs.

The main pressures are on the annual MCI targets and means and on the way of calculating the standard unit cost, especially over the distribution of the manufacturing overhead at the product level. Annual MCI targets will target both the improvement of transformation costs (especially behind *time-related losses*—TRL) and material costs (especially behind *physical losses*—PL and waste).

Therefore, from the perspective of the MCPD system, the annual MCI goal level of current and future products represents the point of interest. Planning and tracking the effectiveness and efficiency of kaizen and kaikaku projects and daily cost management are the key points of the MCPD system. Annual MCI target becomes unnecessary without the timely completion of annual MCI means targets.

The second simulation method starts from the target profit of a representative product of a PFC, usually the one with the highest pressure on the annual MCI goal, and continues by increasing or decreasing the manufacturing target profit of the other individual products of the PFC, depending on the volume of target sales, the evolution of the target price, and the need for internal profit. Therefore, the target profit determination is built around a representative product that usually determines the annual MCI target level and, implicitly, the MCI targets and means required to be achieved for a PFC. The simulation has the same ingredients as the previous one.

In fact, through these simulations, no matter which method is used, the annual MCI goal and annual MCI targets and means are determined for each PFC and/or product as the difference between the annual manufacturing target profit and the annual external manufacturing profit estimated, considering the evolution of the sales volumes, prices, and opportunities for MCI, to meet the company's long-term profit plan (multiannual target profit). Sometimes, compensations can be established between the target profits of two or more PFCs or between PFC component products at different times in their life cycle.

However, a major point of interest of the MCPD system is the cost system accuracy, with an extension to the budgeting system.

3.1.2.2 Cost System Analysis and the Needs for an MCPD System

The MCPD system is not a costing system. This is a system that directs improvements and transformation of manufacturing flow for each PFC to meet the need for MCI imposed by market competitiveness and by the need for multiannual profit. From the MCPD system perspective, the manufacturing cost performance analysis is carried out by monitoring and continuously analyzing the annual MCI goal and the annual MCI targets variations for each PFC and the current state of the manufacturing costs (the level of reaching targets). The concrete results of meeting the annual MCI targets should be visible in reducing the unit manufacturing cost. This is the approach to the annual manufacturing improvement budget and internal profit (annual MCI goal). For the analysis of the external profit performance (obtained from sales; annual external profit), the traditional analysis of the budgetary variations and the underlying standard cost changes continues.

Considering that in manufacturing companies, irrespective of the working environments (*Computer Numerically Controlled, Computer Integrated Manufacturing, or Industry 4.0*), the number of hours of equipment running is often much higher than that of direct labor hours, the continuous measurement of the *value-adding operating time* (*VAOT*) of equipment and all *equipment working hours* (*EWH*) associated costs, including the cost of utilities consumed by equipment, is a priority. Therefore, calculating a minute of equipment and/or human activity and determining the unnecessary consumption of materials and utilities related to non-value-added activities can be achieved using various well-established costing methods, such as *Standard Costing* (*SC*) and *Activity-Based Costing* (*ABC*), for assigning manufacturing overhead costs to products. Using the costing systems, the MCPD system tracks the calculation and reduction of the CLW for each PFC process in order to identify the non-value-added cost reserve from the unit manufacturing cost structure. Therefore, the processing time, which is the time it takes for a product to go through the entire manufacturing flow, is very important because it must tend to the level of the theoretical capacity. In fact, it is intended to achieve the highest possible level of OEE with the help of kaizen projects and a synchronization of the cycle time from the bottleneck process in comparison to the maximum takt time to meet customer demands. In this way, the reduction and/or elimination of the level of losses and waste and, implicitly, the costs behind them is pursued.

From the perspective of the MCPD system, the manufacturing costs structure is as follows:

■ Transformation costs
  – Direct labor costs (variable or fixed costs; depends on the company organization)
  – Indirect labor costs (variable costs)
  – Manufacturing overhead (variable costs, in general)
    • Maintenance costs/spare parts cost (variable or fixed costs, depends on the company organization)
    • Utility costs (variable costs)
    • Tool costs (variable costs)
    • Die and jig costs (variable costs)
  – Depreciation costs (fixed costs)
■ Material costs (variable costs)
  – Direct material costs
  – Indirect material costs (auxiliary materials cost)

The general rule of the MCPD system is to ensure that as much of the indirect costs as possible become direct (allocating as many manufacturing costs as possible to processes level and then to product level). The variable indirect costs are those that hide most of the losses and waste. However, at the level of each cost structure mentioned earlier, losses and waste are hidden in varying percentages. The purpose of the MCPD system is to discover them and to reduce and/or eliminate them through the annual MCI means.

Selling, general, and administrative expenses are not part of the MCPD system area.

CLWs that cannot be removed for the moment are treated as an acceptable part of current costs.

Therefore, in order to support MCI targets and means annually, a direct and more complete allocation of cost-to-cost centers is sought, such as maintenance cost, auxiliary materials cost, utilities cost, depreciation cost, cost of transportation between internal warehouses, quality testing cost, and external services (maintenance, ISO certifications, consulting, training, cleaning, security, quality tests, etc.). Exceptions are only rent costs, insurance costs, and taxes costs, which are further treated as indirect (fixed) and assigned to unit cost manufacturing by *predetermined* percentages of transformation costs at each cost center or physical area.

The continuous increase of VAOT through kaizen projects results in the distribution of reduced manufacturing overhead costs values remaining at each cost center, which leads to the knowledge and continuous reduction of the unit manufacturing costs for each PFC, regardless of whether the existing products and/or future products are targeted.

Manufacturing costs are assigned to cost centers, which are often work cells or islands. The calculation method on manufacturing phases is used to determine the unit manufacturing costs.

From the MCPD system perspective, it is important first to know the level of cost transformation and material costs at the level of each process (cost center; usually a piece of equipment) and then to know the level of CLW at the level of each process. In this way, it is possible to set pertinent targets to reduce losses and waste and, implicitly, CLW. CLW is obtained by transforming into manufacturing costs a: (1) non-value-added minutes of equipment, (2) non-value-added minutes of human work, and (3) unnecessary use of utilities, materials and stocks.

In this context of the MCPD system, besides the fact that as many costs as possible are directly attributed to the processes, the potential level of the annual MCI target at the level of processes and activities and at the level of each PFC is also known. Having in mind the reduction and/or elimination of CLW through kaizen and kaikaku projects, the annual MCI goal and annual MCI targets and means become a continuous strategic activity. In this context, at product level, the analysis of the variations between the current CLW target and the CLW and CCLW associated targets levels becomes a more important activity from the perspective of cost/price and profit competitiveness than the analysis of the variations between standard costs and current costs at the product level. The choice of types of losses and waste is at the discretion of each manufacturing company according to the specifics of their activity and their own needs for a certain period of time. The main goal is to capture as much percentage as possible of the CLW and CCLW from the unit manufacturing structure, and to continually reduce this percentage through productivity and quality improvements (annual MCI means), which are continually directed at the root of acute cost problems (CCLW) and at the root of chronic cost problems (CLW), with a well-known and assumed feasibility of annual improvements.

In conclusion, the cost system should provide a situation of the unit manufacturing costs as accurate as possible on the standard and actual consumption levels for each PFC/product/customer and, implicitly, for each process/work center/activity. At the same time, there is a need for reconciliation, through successive simulations, between the annual manufacturing target profit and the annual MCI targets (MCI targets level 2 reconciliation).

3.1.3 Driven by Annual MCI Targets Deployment

To support the annual MCI goal at the level of each PFC of a company, the annual MCI targets deployment is established at the level of major processes or all the processes involved in each PFC (see Formula 1.5). For this, annual CLW targets and annual CCLW targets are set at the level of each major process or all PFC processes (see Formula 1.6) based on the annual MCI means targets at each PFC approach process (see Formulas 1.8 and 1.9).

Annual MCI targets deployment helps to reduce and/or eliminate the first two major barriers to sustaining the culture of continuous improvement and, implicitly, the consistent transformation of manufacturing companies, presented in Section 1.1.1, respectively: *real managerial commitment and resistance to change.*

Annual MCI targets deployment may aim at establishing annual targets for CCLW and CLW and at product, client, work center, department, and activity levels.

To accomplish the first steps of the MCPD system, it is important to understand that MCI target deployment requires two approaches: (1) measurement to prevent and support an acceptable level of annual CLW targets and annual CCLW targets (based on MCI means level 1) and (2) measurement for systemic and systematic improvements of the annual CLW targets and annual CCLW targets (based on MCI means levels 2 and 3). Annual CLW measurement and annual CCLW require first to locate as accurately as possible the level of processes and flow of each PFC of losses and waste. From the perspective of CCLW or TRL, the continuous measurement of the real time available for achieving the production and the actual differences between the standard cycle time and the actual cycle time for each operation are still a challenge for many companies. Not often, companies are planning far beyond normal production capacities, resulting in losses and waste and, implicitly, CLW. In this context, the continuous measurement of current capabilities and current WIP level and the promotion of problem-solving and systematic and systemic improvement activities can lead to a scientific and consistent reduction of the unit manufacturing costs and beyond.

3.1.3.1 Understanding the Manufacturing Flow from the Perspective of an MCPD System

At the same time or after accurately confirming the unit manufacturing cost level, the losses and waste are addressed by deeply understanding the principles and phenomena that occur at the manufacturing flow level.

Manufacturing flow is just a part of the entire delivery flow of an order to the customer (*total lead time—TLT*). For example, the TLT may last for 75 days, which, in addition to the MLT of 210.5 hours or 3.5 days (with WIP of 202.08 hours and setup time together with a processing time of 8.42 hours), includes the times for the following processes: order processing, planning, preparation of raw material, transfer to the finished products warehouse, and finished products warehouse and truck loading.

The understanding of the manufacturing flow by managers in terms of MCPD system implications aims at locating, measuring, and continuously reducing CLW for the shop floor processes (manufacturing/man and maintenance/equipment) in order to build a stable and low-cost work environment and a system that continuously synchronizes to market signals (product quantities, cost, quality, delivery time) through a harmonious transformation of processes to which all people in and beyond the company can participate.

But, in order to address CLW, it is necessary to have knowledge of the losses and waste. Both measurement to prevent and support an acceptable level of CLW and CCLW and the measurement of losses and waste for systematic and systemic improvements require a thorough knowledge of all technological processes, of all technological process operations, of all phases of technological operations, and of all activities of technological operations. Furthermore, for all technological and not only technological operational activities, it is necessary to determine all the constraints that affect the current unit manufacturing cost level, that is, constraints that lead to a part of the manufacturing costs being non-value-added costs (hidden behind losses and waste).

To understand and interrogate on a continuous basis the manufacturing flow in order to identify the annual MCI means, specifically the four destinations previously described in Section 2.4.1.4, to help meet the annual MCI targets (by meeting the annual CLW targets and the annual CCLW targets) imposed by the annual MCI goal, the following are pursued:

1. The continuous reduction of *MLT* by seeking implementation in as many areas of the manufacturing flow as possible, if not all, of the *continuous flow* (*CF*) and *OPF* principles
2. Synchronization of all manufacturing processes at the *takt time*
3. Reduction of *WIP* stocks
4. Reducing the number of *changes in production planning*
5. *Overtime* reduction and/or elimination
6. Increasing the accuracy of *internal logistics capacity planning* (material handling—type, frequencies, duration, and distances)
7. Reduction of stocks of *raw materials and components*

Understanding manufacturing flow from the MCPD system perspective is achieved for two levels of thinking, which are discussed in the next sections.

3.1.3.1.1 Focus on Demand and Capacity: Measurement of Present Capacity

Ohno, when defining the seven main types of waste (Ohno, 1988, pp. 19–20), is relying on the following equation: "Present capacity = work + waste. True efficiency improvement comes when we produce zero waste and bring the percentage of work to 100 percent." Starting from this Ohno's concept, one can say that the current capacity for each manufacturing process is as follows:

$$\text{Present capacity} = \text{Cycle time} + \text{Losses} \tag{3.1}$$

Note 1: Losses mean the 16 main types of losses developed by Nakajima (1988).

Note 2: The concept of losses is used instead of the waste concept as it is considered to be more comprehensive than the last one; the concept of waste in the MCPD vision targets stocks (Posteucă and Sakamoto, 2017, p. 29).

At the same time, the present capacity for each manufacturing process can be expressed as such (Tapping et al., 2002, p. 113):

Present capacity = Available production time/Cycle time for the operation

(3.2)

Analyzing the earlier two formulas from the manufacturing perspective, it can be said that the current capacity level is influenced by the stability and the level of OEE (including setup time), which affects the available production time and the variations between the actual cycle time and the standard cycle time (as the size of the OEE). More precisely, the lower the capacity level, the losses level being high, the available production time is lower, and the unit manufacturing cost structure incorporates a high level of CLW (unit manufacturing costs with multiple improvement opportunities). Moreover, the higher the actual cycle time is than the standard cycle time, also as losses (performance-related losses), the lower the capacity level, and, implicitly, the unit manufacturing cost structure incorporates a high level of CLW that is associated with performance losses.

3.1.3.1.2 Focus on Manufacturing Flow: Ideal State vs. Current State

Moving forward, at the manufacturing flow level, the deviation between the ideal and the current state of the losses and waste levels is determined for each PFC. The ideal state of the manufacturing flow is the state where the *OPF* principle is found at all processes.

As it is known, MLT can be determined as such (Womack and Jones, 2011; Tapping et al., 2002):

Manufacturing lead time = Total cycle time + Waste   (3.3)

Note 1: Waste (Ohno, 1978, 1982, 1988) means WIP.

MLT could be in the ideal state, or 100% work according to Ohno's formula; *if the OPF principle* would be totally achieved (Sekine, 2005), more precisely, each cycle time would fall into the takt time without losses (with a larger OEE) and only the minimum time related to the work-indispensable activities (including setup time and material transfer time) would be accepted, then MLT would be

$$\text{Manufacturing lead time} = \text{TCT} + \text{WIP}(S) + \text{WIP}(T1) \qquad (3.4)$$

where

TCT is the *total cycle time* (the sum of all cycle times that are perfectly balanced at the takt time—measured in seconds; considering that there is no bottleneck)

WIP (S) is the setup-related WIP (the impact on WIP, in number of stocks, of all setup activities of the various processes of the manufacturing flow requiring setup activities upon batch change; under the conditions of exact compliance with the standard setup time, possibly less than 10 minutes each)

WIP (T1) is the WIP related to the transfer activities (it is the impact on WIP, in number of stocks, of the ongoing stockpile transfer activities, beyond the cycle time from the previous process to the next process, moment when it is immediately taken over by the cycle time of the next process; under the strict observance of the standard transfer time, usually automatic transfers by elevators, rotating tables, etc.)

In this context, both WIP resulting from setup activities and WIP resulting from stock transfer activities beyond the cycle time and up to the next process could be assimilated to Ohno's "work," and work would become 100% and would meet the concepts of "true efficiency" and "zero waste" (Ohno, 1988, pp. 19–20).

However, the current situation from the manufacturing companies differs from the ideal situation mentioned earlier (Formula 3.4), and the following phenomena often occur:

■ *Cycle time non-synchronized to the takt time* causes the occurrence of WIP related to the transfer between the processes within the manufacturing flow, especially if the cycle time in *bottleneck* processes/activities is not synchronized to the maximum takt time required by the customer requests.

■ *The WIP level as a result of the setup* often varies because the minimal and maximum planning of this WIP sometimes does not take into account the standard setup times of each process, the current variations of the setup standard, and the production mix planning.

■ *The level of WIP resulting from the transfer of stocks beyond the cycle time and up to the next process* varies from multiple possible causes such as operator skills differences, job ergonomics, method of making the transfer, etc. For example, a breakdown equipment will immediately generate WIP both in the process where the equipment breakdown occurred and on the total manufacturing flow—it is one of the forms of CCLW occurrence.

■ *The entire WIP level* (resulting from setup activities, stock transfer activities beyond cycle time and before moving to the process, transferring to the next process until cycle time starts, or setup activities in the next process) is under the influence of the MCPD principle of "waste (stocks) elasticity on losses" (Posteucă and Sakamoto, 2017, pp. 85, 236–238). This principle expresses the stock variation (stock level of raw materials, components, and WIP) usually negative (increase in stocks) due to the change in the standard losses level (equipment, people, material, and utilities) for which production capacity was originally planned to meet customers' demands.

Considering these phenomena, the MLT becomes as follows:

$$\text{Manufacturing lead time} = \text{TCT} + \text{WIP}\left(\text{S}\right) + \text{WIP}\left(\text{T2}\right) \qquad (3.5)$$

where, besides the previous explanations, we have WIP (T2) as the WIP related to all transfer activities between the manufacturing flow processes (the impact on WIP of the transfer activities from one process to another, which is often affected by the present capacity level of the previous process and the following process—by current constraints).

In this context of Formula 3.5, from the perspective of the MCPD system, CLW appear both within the cycle time of processes involving people and/or equipment (especially in bottleneck processes/activities and before and beyond bottleneck), in WIP related to setup activities, and especially in the rest of the activities that determine the remainder of the WIP (capacity level affected by a low level of OEE). Therefore, the annual MCI targets deployment will aim at identifying the annual MCI means for cycle time, for setup-related WIP, and the remaining WIP affected by the low level of OEE.

For example, in order to address the reduction of the entire WIP, the best ways to know the WIP level at any time will be searched for, to take into account the planning of the production mix, considering how WIP is created (the variation of process constraints), designing assistive devices to ensure fast transfer between processes of stocks under processing, and considering the need to reduce the batch size to increase flexibility and to set a minimum and maximum level for each type of stock, to continuously increase OEE, etc.

Therefore, understanding the manufacturing flow from the perspective of the MCPD system requires the following:

■ *Providing a framework for setting up the annual MCI targets deployment* (reducing materials costs by triggering in-depth cost study on value-added analysis—costs of PL and costs of waste; reducing transformation costs by reducing or eliminating non-value-added costs from processes, costs of TRL; especially those causing other losses and waste along the manufacturing flow, CCLW)
■ *Continuous identification of the annual MCI means* (providing concrete means to continually fulfill the annual MCI targets by *improving the MLT*) by addressing the following at the level of each process: (1) *productivity* (reducing or eliminating non-value-added activities and the lead time of each manufacturing process), (2) *capacity* (planning people's work and equipment to omit worthless work), and (3) *product quality ratio* (reducing MLT by reducing non-quality incidents)

In conclusion, by continually comparing MLT in the ideal state (Formula 3.4) with the current situation (Formula 3.5—measured with the best accuracy), the direction of the MCPD system and, implicitly, the establishment of annual MCI targets and means are determined on a continuous basis.

For example, CC-Plant manufactures engine components for the automotive industry and has 10 successive product-processing processes ("A") within PFC1. In the first process ("P1"), there are four basic activities ("a1," "a2," "a3," and "a4"). By comparing Formulas 3.4 and 3.5, the future state of the MLT for the "P1" is identified to meet the annual MCI goal at the PFC1 level.

Therefore, according to Formula 3.4 (ideal state), we have

Manufacturing lead time for process "P1" $= TCT + WIP(S) + WIP(T1)$

$$= 16.2\,min + 9\,min + 6\,min = 31.2\,min$$

where
TCT of 16.2 min = 0.5 min in "a1" + 5.5 min in "a2" + 4.7 min in "a3" + 5.5 min in "a4"
WIP (S) of 9 min = 9 min of "a1"
    Note: In "a2" and "a3," there is a setup of 2 minutes each, but concomitantly with setup of "a1."
WIP (T1) of 6 min = 1 min of "a1" to "a2" + 0.5 min of "a2" to "a3" + 0.5 min of "a3" to "a4" + 4.0 min of "a4" at process "P2"

But according to Formula 3.5, the current state of the MLT is

Manufacturing lead time for process "P1" $= TCT + WIP(S) + WIP(T2)$

$$= 18.2 \min + 17 \min + 29 \min = 64.2 \min$$

where

TCT of 18.2 min = 0.5 min in "a1" + 5.5 min in "a2" + 5.5 min in "a3" + 6.5 min in "a4"

Note 1: Activity "a3" has a deviation of the average cycle time of 1 min (from 4.7 min standard cycle time to 5.7 min actual cycle time).

Note 2: Activity "a4" has a deviation of the cycle time of 1 min (from 5.5 min standard cycle time to 6.5 min actual cycle time).

WIP (S) of 17 min = 17 min of "a1"

Note 3: In "a2" setup time is of 8 min and "a3" setup time is of 7 min but is performed concomitantly with the setup of "a1."

WIP (T2) of 29 min is due to the lack of synchronization between the "P1" and the "P2" amid the capacity variations in the two processes, including the cycle time and beyond (e.g., breakdown time, start-up time, minor stoppage/idling time, defects and rework time, etc.)

Therefore, in order to reduce the 33-minute difference between the two MLT situations for "P1" (64.2 min – 31.2 min), improvement targets are set in order to reduce and/or eliminate all the elements contributing to the 33-minute deviation, starting with the stabilizing and increasing current capacities (reducing and/or eliminating losses that have an impact on WIP), especially bottleneck activities that have to cope with the maximum level of the takt time needed. Behind the 33 minutes of "P1," there are costs that lead to the formation of a unit manufacturing cost for "P1" that incorporates in its structure cost elements that have no added value. Moreover, some of the "P1" activities can be bottleneck activities for the entire manufacturing flow, generating non-value-added times and unnecessary use of materials and utilities and, implicitly, unnecessary costs throughout the entire manufacturing flow. Similarly, all the processes where raw materials are transformed into finished product "A" are addressed to identify all the opportunities for achieving the annual MCI targets that must converge to the annual MCI goal of each PFC. In this way, the continuous beneficial transformation of the manufacturing flow is achieved in order to ensure a long-term capacity that satisfies the productivity vision and mission (the volumes of products manufactured in order to sell them) and a competitive cost level.

Understanding the manufacturing flow from the perspective of an MCPD system is required for both existing and future products. The main goal is to ensure the level of annual and multiannual target profit.

3.1.3.2 Losses and Waste and Critical Losses and Waste for Each PFC

In order to confirm the full understanding of the MLT from the MCPD system perspective, it is necessary to locate and measure continuously the losses and waste of each manufacturing process and critical losses and waste (root cause of losses and waste along the manufacturing flow).

From the MCI perspective, the purpose of any manufacturing company is to continuously improve the efficiency and effectiveness of processes by maximizing outputs and minimizing inputs. In order to reduce and/or eliminate losses that impact WIP (stocks as effects), it is necessary that all the inputs and outputs of each PFC process are analyzed in order to identify all reduction opportunities for the time that do not add value and all unnecessary use of materials and utilities.

The basic inputs of any manufacturing process based on which outputs are produced in the form of products/services/information are as follows: *raw materials and components, consumables* (lubricants, coolants, hydraulic fluids, drills, etc.), *information* [work instructions, article, quantity, delivery deadline; knowledge and skills, theoretical knowledge, practical skills; Standard Operating Procedure (SOP), procedures methods], *environment elements* (cleanliness, brightness, vibration, noise, temperature, humidity), *machinery* (dies, jigs, tools, gauges), and *utilities* (electricity, gas, steam, compressed air, water, etc.). The manufacturing cost level is affected by these main resource inputs in each manufacturing process and, implicitly, in each output of each PFC.

Increased efficiency is achieved by minimizing inputs (excess amount of input or waste or too many stocks due to the fluctuating capacity of the equipment in particular) and maintaining outputs at least at the same level. Increased effectiveness is achieved by maximizing outputs (not effectively used input or losses or too much time consumed—of equipment, human works, and consumption of raw materials, materials, and utilities) and maintaining inputs at the same level.

Determining losses and waste involves establishing *weaknesses* in each process/work center of the manufacturing flow, while critical losses and waste involves establishing *MKP* by identifying the causal relationships between losses and waste considered critical and the other losses and waste resulted. Determining critical losses and waste helps to set priorities for annual MCI targets, especially for TRL. Continuous measurement of losses and waste (KPIs of losses and waste) is the key challenge in setting the annual MCI targets and means (Posteucă and Sakamoto, 2017, pp. 119–122). This measurement takes place continuously from shift to shift. Establishing critical losses and waste is done by analyzing all nonproductive times of a process or equipment, from shift to shift, by measuring all the effects of each loss and waste, considered as the root cause for other losses and waste, effects that are visible by forcibly stopping or reducing the activity of a process or equipment. When a process is stopped or the activity is reduced, the "5 Whys?" technique and real-time brainstorming are used until the root cause and the type of losses and waste are identified (e.g., stopping equipment from lack of raw material, including noncompliant raw material). The root cause of this shutdown is not the equipment that will record a lower percentage of OEE, including a higher level of CLW, but other processes/work centers/activities. Figure 3.1 presents the causal relationship between losses and waste for equipment lacking raw material at a given time and potential root causes. Such root causes are

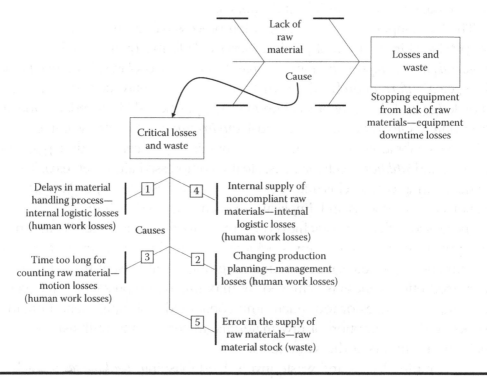

Figure 3.1　Critical losses and waste: cause–effect relationships.

monitored daily, being the potential causes of losses and waste for other equipment as well (processes). The stake in identifying such root cause, such critical losses and waste is to reduce the missed opportunities of achieving the planned production, implicitly increasing WIP, and increasing CCLW and CLW.

At the end of the month, there is a calculation of the critical losses and waste and a related Pareto graph (in the example given earlier, the five causes are hierarchized). Subsequently, they will be the subject of annual MCI means deployment if the CCLW level is significant and if they were chosen to be reduced to meet the annual MCI goal. This way, the total losses and waste of each process will be reduced by calculating the critical losses and waste (some of the losses and waste from a process will be attributed to critical losses and waste from other processes). The remaining losses and waste will be subject to annual MCI means if the CLW level is significant and if they were chosen to be reduced to meet the annual MCI goal. Therefore, some processes/work centers will only have to improve their own losses and waste, which do not have extensive effects, while others, besides their own (isolated) losses and waste, will have to improve the critical losses and waste as well (contamination of personal losses and waste on other processes or other equipment). Critical losses and waste usually have a priority in approach as they are the current acute problems. However, through the subsequent transformation of critical losses and waste into costs (CCLW) and losses and waste into costs (CLW), priorities will be set so as to achieve annual MCI targets through annual MCI means to meet the annual MCI goal at the level of each PFC.

Therefore, for every process/work center (cost center), such as equipment, it is necessary to constantly know the TRL and PL related to each type of losses and waste. For example, for continuous knowledge of breakdown losses (downtime losses), the following will be measured: (1) breakdown rate of equipment (1/hour), (2) repair rate of equipment (1/hour), and (3) number of defects produced by the breakdown (unit) (Posteucă and Sakamoto, 2017, pp. 119–122). Moreover, the *critical losses and waste* related to the processes and activities affected by breakdown upstream and downstream of the entire manufacturing flow for each PFC will be determined. Then, hierarchies of the impact of each type of losses/waste on each process will be established, and also the type of losses and waste that represents the main cause and the impact on non-value-added time (*TRL, equipment and human work*) and unnecessarily consumed materials and utilities (*PL, materials and energy*) along the entire manufacturing flow. In this way, the annual MCI means will be targeted toward meeting the annual MCI targets.

The losses level for an equipment is determined with the aid of OEE. OEE is determined for equipment considered important, especially for bottleneck equipment, knowing the losses for each: part number, customer, shift, department, work center, and PFC.

For example, Table 3.1 presents the OEE calculation (monthly average) based on data collected over six months for thermoforming machine.

Extending the abovementioned example for the thermoforming machine, Table 3.2 shows the percentage on every loss from the equipment loading time (ELT) for thermoforming machine.

As may be seen, the equipment capacity is high (OEE is of 91.3%, but its flexibility is relatively low). Therefore, the opportunity to improve flexibility by reducing the setup time is obvious, especially since setup loss can be critical losses and waste for other processes.

Furthermore, Figure 3.2 plots the capacity for the thermoforming machine. As one can see, the level of OEE is high compared to the World Class of 85% (availability percentage of 90.0%, performance percentage of 95%, and quality percentage of 99.5%). In this case, for the thermoforming machine, several kaizen projects have been carried out over the lifetime of the equipment, and in the next period, it is envisaged to change the equipment with a new one, in order to meet the demand for the additional capacity imposed by the customers' demand (systemic improvement project is necessary—kaikaku; MCI means level 3).

If the type of industry is fabrication and assembly machinery (automotive, metal products, appliances, electrical products, precision instruments, etc.), the level losses is calculated using the *Overall Line Effectiveness* (*OLE*) to capture the level of synchronization between the assembly line equipment. In order to determine the OLE, it is necessary that the takt time for the selected processes be the same. The OEE calculation can perform as a product of the OEE indicators from the selected processes. For example, Table 3.3 calculates OLE for four processes that have the same takt time. As one can see, OLE is 0.370 or 37%.

In fact, OLE values will tend more and more toward "0" (zero) as the number of equipment monitored in the line increases and the *quality* level % will decrease as the last process in the line is reached, because the quality issues present in each previous process in the line affect the following processes. However, OLE's basic purpose, as well as the purpose of OEE, is to identify itself rapidly and to continually steer improvement opportunities through a calculation simple to perform. For this, it is necessary to continuously maintain the OEE and OLE calculation method and to collect all data carefully for comparative analysis over time.

Table 3.1 OEE Calculation

	Calculation of Losses of Thermoforming Machines	Monthly Average (min)	No. of Events/ Month	Average Time/ Event (min)	Scrap
1.	Equipment working hours (EWH) (30 days * 8 hours * 60 minutes)	14,400	×	×	×
2.	Equipment scheduled downtime (ESD) (a + b)	1,380	×	×	×
a.	Admissible stops caused by equipment (ASE) (a1 + a2 + a3)	220	×	×	×
a1.	Time for cleaning	150	50	3	0
a2.	Time for lubricating	60	4	15	0
a3.	Time for planned maintenance	10	1	10	0
b.	Independent stops caused by equipment considion (ISE) (b1 + b2 + b3 + b4 + b5 + b6)	1,160	×	×	×
b1.	Lunch break (maximum 30 minutes)	900	30	30	0
b2.	Short breaks (5–10 minutes/break)	150	30	5	0
b3.	Time for training	60	2	30	0
b4.	Time for lack of tasks	25	5	5	0
b5.	Electricity interruption time	10	1	10	1
b6.	Waiting for materials	15	3	5	0
3.	Equipment loading time [ELT] (1) − (2)	13,020	×	×	×
4.	Breakdown time (b1) (c + d)	35	×	×	×
c.	Tools failure (mold)	25	1	25	1
d.	Waiting for repairs	10	1	10	0

(Continued)

Table 3.1 (Continued)   OEE Calculation

	Calculation of Losses of Thermoforming Machines	Monthly Average (min)	No. of Events/ Month	Average Time/ Event (min)	Scrap
5.	Setup and adjustment time	638	22	29	14
6.	Cutting tool replacement time	15	5	3	0
7.	Start-up and yield	50	5	10	10
8.	Equipment utilization time (EUT) (3) − (4) − (5) − (6) − (7)	12,282	×	×	×
9.	Loss of speed [Ls][a]	290	×	×	×
10.	Minor stops and idling [MSI][b]	77	×	×	×
11.	Equipment net utilization time (ENUT) (Sct * N)	11,924	×	×	×
12.	Rework time	25	25	1	×
13.	Total loss with scrap [TLS][c]	17	×	×	×
14.	Value-adding operating time [VAOT][d]	11,872.9	×	×	×

OEE = Availability × Performance × Quality = EUT/ELT × ENUT/EUT × VAOT/ENUT = 12,282/13,020 × 11,924/12,282 × 11,872.9/11,924

$\quad$ = 0.943 × 0.970 × 0.995 = 0.91

or

OEE = VAOT/ELT = 11,872.9/13,020 = 0.91.

Notes:
[a] Ls = N * Rct-Sct = 17,400 pieces * (42 sec − 41 sec) = 17,400 * 1 sec = 17,400 sec = 290 min.
[b] MSI = EWH − (ESD + 4 + 5 + 6 + 7 + 9 + 12 + 13) − VAOT = 14,400 − (1,380 + 35 + 638 + 15 + 50 + 290 + 25 + 17) − 11,872.9 = 14,400 − 2,450 − 11,872.9 = 77 min.
[c] TLS = Sct * S = 41 sec/piece * 25 pieces = 1,025 sec = 17 min.
[d] VAOT = [Sct*(N-S)] = 41 sec/piece * (17,400 pieces − 25 pieces) = 11,872.9 min.

Table 3.2 Losses Structure and OEE in ELT

	%	Minutes/Month
OEE	91.2	11,890
Loss of speed	2.2	290
Breakdown	0.3	35
Setup	4.9	638
Cutting tool	0.1	15
Minor stops	0.6	77
Start-up	0.4	50
Rework + scrap	0.3	42
Equipment loading time (ELT)	100%	13,020

Therefore, following the continuous measurement of OEE for equipment chosen as strategic, especially for bottleneck ones, the actual capacity of a piece of equipment (declared as process and cost center) can be determined and WIP (S) can be measured. Furthermore, by adding WIP (T2), the current MLT for a process is determined. To determine the MLT for each PFC, the MLT for each process is summed up. By comparing continuously WIP (T1) with WIP (T2) for each PFC, Formulas 3.4 and 3.5 serve to determine the *critical WIP points* underlying the deviation between WIP (T2) and WIP (T1). The goal is to continually reduce this variation between the WIP's ideal state and WIP's current state by continuously designing a future WIP state that is achieved through kaizen and kaikaku projects (MCI means levels 2 and 3) for each PFC, *to reduce the waste level continuously.*

For this, besides analyses to improve the takt time, of line balancing, the man*hour, the number of operations, the bottleneck, the distances traveled, and the setup times, an analysis is performed especially for the following:

1. *Transfer times* (optimizing transfer times, eliminating delays in mechanical transfers—elevators, rotating tables, etc., reducing and/or eliminating cumbersome manual manipulations, the possibility of maximum use of automatic transfers, etc.)
2. *WIP management* (designing a WIP ordering system, WIP map definition, WIP definition by setup time, production mix, production plan changes, equipment capacity constraints, continuous WIP monitoring during processing, continually defining and redefining the minimum and maximum WIP levels, etc.)

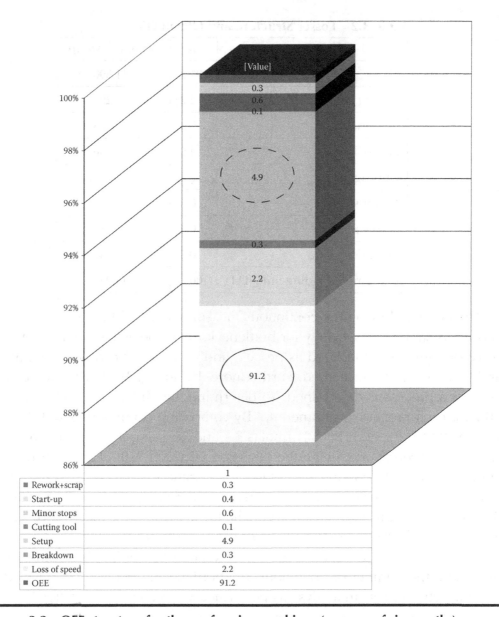

Figure 3.2 OEE structure for thermoforming machines (average of six months).

The design of WIP's future state between the manufacturing processes will start especially from *critical waste*, which represents WIP constraints in terms of *setup, transfer times,* and *WIP management.* These constraints affect the entire manufacturing flow of a PFC. Designing the future state of process capacities (especially for the equipment that provide preponderantly speed) will start particularly from *critical losses* that differ from one process to another, from one equipment to another, and from one period to the

Table 3.3 Overall Line Effectiveness

	OEE in Process 1	OEE in Process 2	OEE in Process 3	OEE in Process 4
Availability %	0.878	0.873	0.856	0.867
Performance %	0.921	0.943	0.956	0.967
Quality %	0.967	0.956	0.941	0.932
OEE	0.7820	0.7870	0.7701	0.7814
OLE = 0.370 (0.7448 * 0.7788 * 0.7627 * 0.8107)				

next, but which must ensure the classification of the maximum takt time in cycle time related to the bottleneck process.

By addressing *critical losses and waste*, taking continuous account of the ideal state of the manufacturing flow (Formula 3.4), an effective transformation of the MLT can be achieved, namely, the consistent and sustainable improvement thereof.

3.1.3.3 Critical Cost of Losses and Waste for Each PFC

In order to continuously target improvements through the need for an annual MCI goal, it is necessary to identify the losses and waste connections and costs associated to *minutes related to losses and waste* (variable and fixed costs of transformation costs for each type of losses—equipment, human work and material, and energy—and each type of waste, WIP and for stocks of raw material/components and finished products) and of *useless consumptions of materials and utilities associated to losses and waste* (variable costs of material costs) at the level of each manufacturing process at each PFC (Posteucă and Sakamoto, 2017, pp. 132–140). This determines the CLW for each manufacturing process and then for MLT for each PFC.

Formula 3.6 shows the CLW calculation for a manufacturing process (chronical issues of the manufacturing costs):

$$\text{CLWMP}(\text{P1}) = \text{CEL} + \text{CHWL} + \text{CMUL} + \text{CW} \qquad (3.6)$$

where

CLWMP is the *cost of losses and waste for a manufacturing process* (P1)
CEL is the *costs of equipment losses* (output)
CHWL is the *costs of human work losses* (output)
CMUL is the *costs of material and energy losses* (output)
CW is the *costs of waste* (input)

Furthermore, at the level of MLT for each PFC, CLW becomes (Formula 3.7)

$$\mathrm{CLWMLT} = \mathrm{CLWMP}\left(\mathrm{P1}\right) + \mathrm{CLWMP}\left(\mathrm{P2}\right) + \mathrm{CLWMP}\left(\mathrm{P3}\right) + \cdots + \mathrm{CLWMP}\left(\mathrm{P\text{"}n\text{"}}\right)$$

$$(3.7)$$

where

CLWMLT is the *cost of losses and waste for manufacturing lead time*

CLWMP (P1) is the *cost of losses and waste for a manufacturing process for process* 1

CLWMP (P2) is the *cost of losses and waste for a manufacturing process for process* 2

CLWMP (P3) is the *cost of losses and waste for a manufacturing process for process* 3

CLWMP (P"n") is the *cost of losses and waste for a manufacturing process for process* "n"

or

$$\mathrm{CLWMLT} = \mathrm{CLWMP}\left(\mathrm{P1} + \mathrm{P2} + \mathrm{P3} + \cdots + \mathrm{P\text{"}n\text{"}}\right) \qquad (3.8)$$

Furthermore, taking into account all *critical losses and waste* from each MLT process for each PFC, the processes and activities considered to be those that generate the *CCLW* (*acute problems of the manufacturing costs*) are determined.

The main types of *weaknesses/MKP* that determine the occurrence of *CCLW* upstream and downstream of the entire MLT for each PFC, consistent with Formula 3.4 (the ideal MLT according to the OPF principle), which have an impact on the takt time, especially in terms of maximum takt time, are as follows:

■ *Variation between the actual cycle time and the standard cycle time in the process considered MKP*

■ *Variations in the preestablished capacity* of each process (OEE and OLE variation over the planned limit at the *availability rate*, excluding setup time, *performance rate*, and *quality rate*, with impact on the reduction of the available time planned for manufacturing and, implicitly, reducing the manufacturing volume and increasing the unit manufacturing cost)

- *Variations of the standard setup time* for each process (variation that differs from one batch of products to another for the same PFC process)
- *Variations in the transfer time preset between processes* (increasing MLT and, implicitly, increasing the unit manufacturing cost)

Therefore, *CCLW* for MLT of each PFC is (Formula 3.9) as follows:

$$\text{CCLW} = \text{CLWrc}\left(\sum \Delta\text{SCTB} + \sum \Delta\text{SPC} + \sum \Delta\text{SST} + \sum \Delta\text{STT}\right)$$

$$+\left[\sum \text{CLWap} * \left(\text{P1} + \text{P2} + \cdots \text{P"n"}\right)\right] \tag{3.9}$$

where
CCLW is the *critical of costs of losses and waste* for a MLT of a PFC
CLWrc is the *costs of losses and waste from root cause process*
$\sum\Delta$SCTB is the sum of *variation of standard cycle time of bottleneck* (*cause*)
$\sum\Delta$SPC is the sum of *variation of standard process capacity* (*cause*)
$\sum\Delta$SST is the sum of *variation of standard setup time* (*cause*)
$\sum\Delta$STT is the sum of *variation of standard transfer time* (*cause*)
$\sum$CLWap is the sum of *costs of losses and waste of the affected processes* (the processes affected by the standard variations) (*effects*)
P1 + P2 + $\cdots$P"n" are the processes of manufacturing flow for a PFC

Therefore, the basic principle of CCLW is that of causality (the relationship between causes and effects). The variation between standard and current values on bottleneck/maximum takt time, process capacity, setup time, and transfer time from a specific process represents the causes of the effects on chain CLW, over upstream and downstream processes and manufacturing flow upstream and beyond the manufacturing flow for a PFC.

In this context, CCLW is a visual tool that captures CLW in dynamics and which can serve to easily identify critical connections between losses, waste, and unit manufacturing costs. The prioritization of the CCLW enhancement approach, as well as the prioritization of CLW, in order to meet the annual MCI targets and, further on, the annual MCI goal at PFC level, is performed depending on the feasibility of opportunities to reduce and/or eliminate losses and/or waste and, implicitly, the weaknesses identified [Θ, high impact of process on CCLW (5 points); Q, some impact of process on CCLW (3 points); Ŏ, limited impact of process on CCLW (1 point)] (see Table 1.3).

In this context, defining and locating each type of losses and waste for each individual process and critical losses and waste with a view to further identifying the relationships between them and the related manufacturing costs represent an important challenge within the MCPD system (see Table 1.3—KPI1, KPI2, ..., KPI7). This determines the total offer for the annual MCI target for each PFC that will meet the ongoing need for unit manufacturing costs reduction imposed by the market requirements.

By pursuing the continuous improvement of CCLW for each PFC, the annual MCI targets focusing area is diminished and all systematic and systemic improvements (annual MCI mean levels 2 and 3) are directed to meet the annual target profit by fulfilling the annual MCI goals and, furthermore, the multiannual targets profit and, implicitly, to uncover the hidden reserves of long-term profitability.

3.1.3.4 From Critical Cost of Losses and Waste to Ideal Cost

The current level of CCLW and CLW for MLT of each PFC provides continuous MCPD system steering and *annual MCI targets and means.*

By continually comparing the current level of CCLW and CLW at the level of each process of each PFC, which represents a percentage of the unit manufacturing cost, with an annual MCI goal, the expected level of the annual MCI targets and means for each PFC is projected. At the same time, by comparing continuously the current level of CCLW and CLW with the ideal level of the MLT, the total opportunity for multiannual internal profit is identified. The cost of any deviation from the ideal level of standard for the cycle time of bottleneck, process capacity, setup time, and transfer time impacting on the entire upstream and downstream manufacturing flow represents *a lost opportunity to achieve the ideal manufacturing costs.* Therefore, in order to reach the *ideal manufacturing cost* state, it is necessary to meet the *zero target of CCLW and zero CLW.* Constantly driven by the ideal level of the manufacturing cost, systematic improvement activities and systemic improvement actions (MCI means levels 2 and 3) are continually aimed at continuously connecting the manufacturing flow to the market, by setting annual MCI targets for actual costs, implicitly for the current level of CCLW and CLW.

Figure 3.3 presents the reducing tensions between target profit and external profit by establishing the internal profit level based on the annual MCI targets and means, the annual CCLW targets, and the annual CLW targets (based on the current level of CLW and CCLW—identified

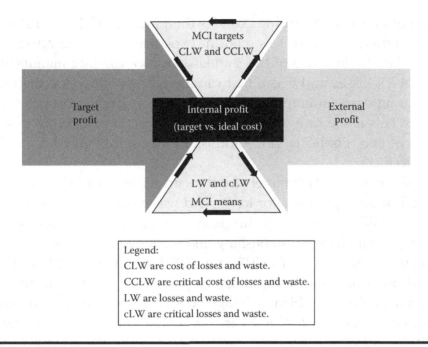

Figure 3.3 Reducing tensions between target profit and external profit by internal profit.

scientifically at the level of processes of each PFC by measuring losses and waste) and taking into account the ideal cost. By continually exploiting the opportunities to improve CCLW and CLW, by setting the annual MCI targets at each process level and by meeting these targets by setting the annual MCI means at the level of each process or system level, one determines the establishing and reaching of unit manufacturing target costs imposed by the market. Establishing the annual MCI targets level is based on the following: (1) the feasibility of the loss and waste approach, located precisely at process level, and (2) the level of manufacturing ideal costs (*zero CCLW and zero CLW* target in the state where the manufacturing flow totally complies with the OPF principle).

The current level of the unit manufacturing costs incorporates a percentage of costs of losses and waste. One of the reasons for exceeding the standard unit manufacturing cost level is to increase the CLW above the expected level. The basic concept of improving the current level of CCLW and CLW at the level of each MLT process of each PFC is to continuously analyze the variation between the current level of CCLW and CLW and the annual MCI targets level, and not to analyze the current state of CCLW and CLW and the standard level for CCLW and CLW. The continuous fulfillment of the annual MCI targets, namely, the continuous reduction of unit manufacturing costs, is the purpose of the MCPD system and not to provide more or less relevant explanations for the causes that contributed to the increase of the standard level of the unit manufacturing costs or which losses and waste had variations with impact on the unit manufacturing costs (attempting to implement consistent solutions, not apologies). This is the only way to achieve consistent competitiveness through cost and, implicitly, price and profit for each PFC and at the level of each product.

Establishing annual MCI targets for the current level of CCLW and CLW is achieved by setting targets for losses and waste detected at each process level of each PFC by improving current operational principles and standards SOP or by radical change. Establishing and meeting the annual MCI targets, based on the current level of CCLW and CLW for MLT of each PFC, is mainly achieved by direct and indirect employees in the manufacturing area, with the help of problem-solving and kaizen projects (continuous setting of new standards for losses and waste and, implicitly, new targets for CCLW and CLW), continuously taking into account the ideal state of the manufacturing flow in which the OPF principle is observed. Continuous identification of solutions steering toward the *ideal costs* and, implicitly, of *zero CCLW and zero CLW targets* through benchmarking studies, the development of scenarios for the complete elimination of all current constraints, and the development of innovative solutions is mainly the task of the technical staff and office staff, through kaizen, but, especially, kaikaku projects (such as innovation of products, equipment, processes, and technologies).

Therefore, all cost problem-solving projects, all kaizen projects, and all kaikaku projects aim at meeting the annual MCI targets and, implicitly, the annual MCI goal based on the offer for the annual MCI targets or, more accurately, the current level of CCLW and CLW for MLT of each PFC (the offer must be at a level of 30%–35% of the total manufacturing cost; this offer is used to choose the annual percentage of the MCI goal to achieve the annual target profit).

In conclusion, the steering of the company's annual harmonious transformation will always consider the annual MCI goal level, the market needs, and the unit manufacturing cost necessary to be achieved to ensure a competitive price and, implicitly, the level of productivity and quality required by the continuous exploitation, in particular of the offer for the current CCLW (the acute cost problems) and subsequently for the CLW (the chronic cost problems), through annual MCI means.

3.1.3.5 Scenario for MCI Targets for Each Product Family

The scenario analysis for each PFC assesses the estimated annual targets level required to be met by MCI at each process level by reducing CCLW and CLW, to meet the annual MCI goal. The basic questions on which the annual MCI targets deployment for each PFC is substantiated are as follows: By how much and where should CCLW and CLW decrease to achieve the annual MCI goal? What is the impact of reducing CCLW and CLW in rising annual and multiannual target sales volumes? What is the impact of reducing CCLW in lowering the target price at product level?

In order to answer these questions and to achieve the annual MCI targets deployment, it is necessary to continuously identify all CLW and CCLW for each manufacturing flow process for each PFC, for both existing and future products. The setting of more marked annual MCI targets for some of the existing products can be made to prepare early on the launch of new profitable products that have to fit into the company's long-term profit plan (multiannual manufacturing target profit). This can happen especially for products that have a long time to market and may be affected by some major cost changes (especially those regarding raw materials and utilities) between the launch of the development plan for the new product and at least during the first months of the launch of the new product, when it is noticed that the level of planned profitability cannot be achieved. In fact, it is necessary to sustain the planned profit (the manufacturing profit) of future products through the internal profit obtained based on the annual MCI goal and, implicitly, by meeting the annual MCI targets (for CCLW and CLW) from processes similar to those of existing products, fulfilling completely the annual MCI means.

For example, for both existing products and future products of a PFC, the current level identified of CCLW and CLW may be of 30%, and the development of the annual MCI target deployment scenario to meet the annual MCI goal can be reducing the unit manufacturing costs by 5.5% for year "N + 1" and 6% for years "N + 2" and "N + 3." Furthermore, to achieve the annual MCI target deployment for the year "N + 1," a total of 3% targets may be established for material and utility costs, and 17.5% for transformation costs for each process and/or work center (see the seven levels of the KPIs in the example in Table 1.3), based on identified and feasible improvement opportunities (annual MCI means).

Furthermore,

■ To meet the 3% reduction target for *material and utility costs*, the following targets for *cost of material and utility losses and costs of waste* (cost of PL) identified at the level of all PFC processes may be established as follows:
   – Costs of direct and indirect labor losses: reduction between 15% and 30% (variable or fixed costs—depends on the company organization)
   – Cost of raw material losses: reduction between 15% and 30% (variable cost losses)
   – Costs of consumables/auxiliary materials losses: reduction between 20% and 30% (variable cost losses)
   – Cost of energy losses: reduction between 20% and 30% (variable cost losses)
   – Costs of die, jig, and tool losses: reduction between 20% and 50% (fixed cost losses)
   – Cost of technological and material scrap losses (yield losses): reduction between 5% and 10% (variable cost losses)
   – Costs of raw material stocks—waste: reduction between 20% and 50% (variable cost losses)
   – Costs of finished products stocks—waste: reduction between 20% and 50% (variable cost losses)
   – Costs of components stocks—waste: reduction between 35% and 50% (variable cost losses)
   – Costs of packaging stocks—waste: reduction between 20% and 35% (variable cost losses)
   – Costs of work in process—waste: reduction between 10% and 15% (variable cost losses)
■ For the 17.5% target for transformation costs, the following targets for transformation cost losses (cost of TRL) identified at the level of all related PFC processes may be established as follows:
   – Costs of scheduled shutdown time losses: reduction between 30% and 50% (variable cost losses)
   – Costs of internal logistic (material handling) losses: reduction between 15% and 30% (variable cost losses)
   – Costs of line organization losses: reduction between 15% and 25% (variable cost losses)
   – Cost of production plan changes losses (management losses): reduction between 15% and 25% (variable cost losses)

- Cost of equipment setup time losses: reduction between 15% and 30% (variable cost losses)
- Cost of equipment breakdown time losses: reduction between 5% and 10% (variable cost losses)
- Cost of equipment speed-down losses: reduction between 25% and 50% (variable cost losses)
- Cost of equipment minor stoppages time losses: reduction between 5% and 10% (variable cost losses)
- Cost of equipment and line rework time losses: reduction between 15% and 25% (variable cost losses)

These targets can be established concurrently with MCI means or not. The earlier percentages are influenced by the current principles and phenomena that determine the current level of CCLW and CLW. The continuous link between the annual MCI targets and annual MCI means determines the best annual MCI goal scenario for which the annual action plan is performed and seeks to involve all employees of the company and all people involved in the MCPD system outside the company. A specific attention is paid to indirect variable cost losses, particularly from the transformation cost losses, to reduce manufacturing overhead costs.

To this end, it is necessary to know and interpret at the level of unit manufacturing costs for each PFC and for the overall company the following information (*planning items for set MCI targets—see Section 1.5*) (*stratification analysis with Pareto graph*):

1. *Transformation and material costs vs. process* (it is usually known; using the costing system)
2. *Process vs. transformation and material costs* (it is usually known; using the costing system)
3. *Losses type/waste type vs. process* (distribution of losses and waste for each process)
4. *Process vs. losses type/waste type* (identified losses/waste structure for each individual process)
5. *Losses type/waste type vs. process step* (distribution of losses and waste for each step of a process)
6. *Process step vs. losses type/waste type* (identified losses/waste structure for each individual process)
7. *CLW vs. process* (CLW determination for each process)
8. *Process vs. CLW* (identified CLW structure for each individual process)

9. *CLW vs. process steps* (CLW determination for each step of a process)
10. *Process steps vs. CLW* (identified CLW structure for each step of a process)
11. *CCLW vs. processes* (determining the CCLW for the process that caused the CCLW and beyond for all processes affected by CCLW by summing all the CLWs at process level)
12. *Process vs. CCLW* (for each and every process it is determined the degree of interference of cost based on CCLW originating from other process(es); to establish the annual MCI targets and means in these processes, the improvement of CCLW shall be established at the source and not where the effects of CCLW are seen, in CLW from other processes)

This way, the scenario for the annual MCI targets deployment for each PFC will be continuously developed by identifying the following information:

■ The structure of total transformation costs and total material and utility costs for each PFC is as follows:
   – Percent of value-added cost
   – Percent of cost of losses and waste
   – Percent cost of losses and waste that can be approached for reduction and/or elimination
   – Percent of CCLW and percent of CLW
   – Percent of CCLW and percent of CLW that are planned for elimination
   – Percent of CCLW and percent of CLW that have already been removed from the beginning of the calendar year or from the beginning of the other reference period
■ The structure of total losses and waste or each PFC is as follows:
   – The total level of each loss and waste
   – Percent possible reduction of each loss and waste through kaizen projects (systematic improvement)
   – Percent possible reduction of each loss and waste through kaikaku projects (systemic improvement)
   – Target for reducing each loss and waste through kaizen projects (systematic improvement)
   – Targets for reducing each loss and waste through kaikaku projects (systemic improvement)

The basic challenge is to identify for each structure the transformation costs and material and utility costs of the actual CLW and CCLW and to continually develop a personalized improvement scenario, as surely each cost structure incorporates a higher or lower percentage of CLW that originates from the process where the resource utilization takes place (attracting costs on cost items) or originating from other processes.

By developing this scenario for the annual MCI targets deployment, the *demand* to reduce the manufacturing costs for each PFC to ensure the level of price competitiveness and annual MCI goal is faced with the *offer* for annual MCI targets, more precisely with the current state of CCLW and CLW. Based on this scenario, the future state of CCLW and CLW, with zero CCLW and zero CLW target, is built by setting the annual MCI targets at process level and process steps to ensure the annual MCI goal directly from the manufacturing process for each PFC, through kaizen and kaikaku projects to increase productivity.

Developing the scenario for the annual MCI targets deployment and prioritizing depending on the feasibility of improvements involve knowing the current state of CCLW and CLW (R, reference) and setting the annual target for MCI (T, targets) and the tracking mechanism of the subsequent evolution of reaching the target established for MCI (TT, % of target touch; 25%, 50%, 75%, and 100%) by the end of the preset reference period (usually the next 12 months with possible six-month adjustments) (see Table 1.3).

At the same time, scenarios for the annual MCI target deployment are developed for a period of one to three years. These scenarios underlie the annual budgeting and, implicitly, the *annual and multiannual improvement budgets development*. This way, annual MCI targets and means will be established, and all necessary resources will be secured on time to sustain, on a continuous basis, a competitive price and an acceptable manufacturing target profit.

The aimed benefits of these multiannual scenarios for the development of the annual MCI targets deployment at processes and process steps level, through the reduction of CCLW and CLW, will be both *tangible* (at the level of increasing production numbers and production capacity, increasing manufacturing profit and unit profit, internal profit growth, manufacturing cost reduction, scrap ratio reduction, rework reduction, MLT reduction, increased production delivery performance [OTIF], etc.) and *intangible* (the management team understands the current, future, and ideal state of manufacturing costs for each PFC, implicitly of the required productivity, increasing teamwork skills at all levels of the company, CCLW and CLW as a visual management tool, etc.).

In order to meet these annual and multiannual scenarios of the MCI targets, reconciliation is needed through successive simulations between MCI targets and MCI means (MCI targets level 3 reconciliation).

3.1.4 Driven by MCI Annual Means Deployment

Starting from the multiannual and annual MCI target deployment scenario, annual MCI means deployment or *action items to set MCI means* (see Section 1.5) are developed for each PFC, then for each department or section and each process/work center/activity of manufacturing flow. At the same time, planning for the action plan (for analyzing, implementing solutions, and monitoring the effects of chosen solutions) is performed. Annual MCI means deployment is the way to continuously promote the true possibilities of addressing the annual CCLW and CLW targets and, implicitly, the productivity, the quality, and sustaining the culture of continuous improvement.

The connection between the three levels of addressing the annual MCI means deployment, each of the three levels based on the Plan–Do–Check–Act Cycle, to meet annual MCI targets and, implicitly, the annual MCI goal, is as follows:

- *Means level 1: Problem-solving techniques (PST) for manufacturing costs*, aim at enhancing the operational control process on the level of CLW targets and on the level of CCLW targets and solving associated problems
- *Means level 2: Kaizen projects to meet MCI targets*, aim to systematically improve the current level of CLW and CCLW through kaizen projects
- *Means level 3: Kaikaku projects to meet MCI targets*, aim at systemic improvement of CLW and CCLW level through kaikaku projects

Priorities in approach are set for the three levels of addressing the MCI means deployment. For *MCI means level 1*, the top five issues are set daily and problem-solving projects are initiated (especially using the A3 technique). For *MCI means level 2* (systemic improvement/*kaizen*) *and MCI means level 3* (systemic improvement/*kaikaku*), the annual strategic projects to improve CLW and CCLW are established. Their prioritization is achieved as follows (see Table 1.3):

- ◎ with five points—high impact of MCI means on MCI targets in the process/system
- ○ with three points—appropriate impact of MCI means on MCI targets in the process/system
- △ with one point—small impact of MCI means on MCI targets in the process/system

Following these priorities, a pertinent number of kaizen and kaikaku projects are set to contribute to meeting the annual MCI goal achieved directly from the manufacturing process for each PFC and reducing the unit manufacturing cost to the level required by the market—constantly taking into consideration the need to achieve the long-term target profit. Therefore, each kaizen and kaikaku project is convergent to the annual MCI goal and, implicitly, to the long-term targets profit.

For each kaizen and kaikaku project, targets for MCI means are set in terms of TRL and/or PL and waste (*MR, means reference* or the initial state of losses and/or waste; *MRT, means-reduction target* or target required to be met for losses and/or waste; and *MTT, % of means target touch,* or tracking the subsequent evolution of target achievement for losses and/or waste and, implicitly, of MCI targets, CLW and CCLW). Annual MCI means deployments (defined at the level of process activities and process level for an entire manufacturing flow of a PFC) are defined so that each member of kaizen or kaikaku projects understands exactly the level of performance pursued and that managers can easily achieve the cost/benefit analysis and determine the level of resources needed to be allocated on time.

Annual MCI means deployment (*action items to set MCI means*) and the achievement on time of the related targets are the true challenge of the MCPD system so that all the work deployed for establishing the annual MCI targets (*planning items to set MCI targets*) is not in vain. However, the exact definition of the problem (annual MCI targets deployment) represents 80% of its solving (annual MCI means deployment). Moreover, as long as annual MCI targets and annual MCI means are established at the same time, with the help of catchball process, through the involvement of all individuals in the company in this reconciliation process, the risk of nonfulfillment of the annual MCI means decreases, implicitly that of the annual MCI goal.

3.1.4.1 Annual Reconciliation to Set MCI Targets: The Basis of the Deployment Mechanism

The stake of the annual reconciliation is to relieve the external environment pressures (*top-down approach*) over the internal environment (*bottom-up approach*) by establishing a consistent approach to the company's transformation in the direction established through the vision and the productivity mission based on a *catchball process.*

The essence of the annual reconciliation is the sizing of the annual external profit and the annual MCI goal so that the annual target profit is met (see Formula 1.2). Uncovering hidden reserves of annual profitability are achieved by meeting the annual MCI goal, which through the driving effects to achieve the annual MCI targets through MCI means has an impact on increasing current capacities and flexibility and reducing manufacturing costs and, implicitly, supporting the annual external profit (from sales). Therefore, meeting the annual MCI targets will target both the annual MCI goal and part of the annual external profit (from sales/manufacturing). The fulfillment of uncovering the hidden reserves of annual profitability is achieved through annual manufacturing flow transformation with the help of the annual MCI means targets (PST, *kaizen* and *kaikaku*) to achieve the annual MCI targets (for CLW and CCLW) that are linked to the annual MCI goal and target price.

In order to achieve the annual reconciliation of MCI targets, it is necessary to involve all managers and specialists of the company. Their goal is to continuously enhance the company's ability to generate the most effective (based on measurements; losses and waste) and efficient (based on cost/benefit analysis) annual MCI means, to meet the annual MCI targets according to their stringent necessity (based on data gathered from the outside of the company and from internal processes, CLW and CCLW) and depending on their ease of accomplishment (people's skills to support the culture of improvement).

In general, setting the annual MCI targets can be done in two ways: *incrementally and comprehensively.* Regardless of the two targeting methods, two questions arise: *What degree of difficulty can the annual MCI target have? How are priorities established in choosing MCI targets to not be perceived as antagonistic?* Responses are based on the author's comments from many production companies. Therefore, the goal of any MCI target is to be met on time and at reasonable cost. Achieving a MCI target will help fulfill the established and necessary performance of the annual MCI goal.

In this context, three directions for establishing MCI targets are possible:

1. A level perceived as low of the annual MCI targets—will discourage increased attention from managers, specialists, and operators
2. An annual MCI target level set by managers themselves, considering the conditions inside the company—will tend to be increasingly smaller and will discourage, sooner or later, the increased attention from all the people involved (will provide many "pertinent" reasons for a low annual MCI target level)
3. A level of annual MCI target perceived as high—can lead to the discouragement of people from the very beginning

Consequently, the answer to the three ways of setting the MCI target is as follows:

a. Annual MCI targets must be connected to the annual MCI goal—to support the acceptable development of the company.
b. Annual MCI targets should be acceptable for those who have to meet them.
c. Annual MCI targets must be perceived as provocative (hard to reach to be motivating).
d. Annual MCI targets must be followed by robust annual MCI means and with a clear and feasible plan—all necessary resources are allocated on time.

From a practical perspective, many companies use incremental target setting for all dimensions of performance (productivity, quality, costs, delivery, etc.). For example, starting from a given current, measured and historical trend situation, the annual productivity growth is set at X% for each of the next five years. To this end, a productivity deployment plan is developed at group level, group company level, department level, compartment level, work centers level, activity level, and, eventually, human level. The problem with this approach is that it is not always possible to set priorities precisely in addressing improvements. Moreover, it continues to aim for *zero losses and waste*, which leads to excessive wear of people and equipment, and this *zero* target is often not linked to the real needs of the market. This approach can support a stressful working environment with a declining level of productivity and quality, which favors high manufacturing costs.

The annual MCI target approach is comprehensive. Annual MCI targets are set at the level of each PFC and/or each product level, according to market needs (top-down approach; sales, price, and profit) and depending on the annual MCI means (bottom-up approach; PST, *kaizen* and *kaikaku* based on CLW and CCLW from process/work centers/activities level). Through comprehensive targets, starting from the current state (standards), the aim is to zero losses and waste, more precisely toward zero cost of losses and waste, with the only difference that targeting and prioritization are scientifically determined. However

■ The expected results are measured in advance in cash.
■ The expected results are accompanied by the individual cost reduction requirements of each type of customer (on each product and/or on each PFC).

- The location of improvement opportunities is accurate (process/work area/activity level).
- The cost/benefit analyses are performed easier and faster (increasing the speed of decision-making on choosing improvements).
- The allocation of resources needed for improvements is easier because the money at stake of any improvement and the contribution of these improvements to the annual MCI goal is known in advance.
- The timing of improvements is easier to achieve because the deadline for price reduction is set (by the customer and/or by the marketing department).
- All managers and people are more easily convinced to continuously participate in the improvements and the continuous and harmonious transformation of the manufacturing company to have an acceptable development in the medium and long term.

The backbone behind the comprehensive MCI targets is to continually have a generous offer on *total costs of losses and waste (TCLW)* and *total critical costs of losses and waste (TCCLW)*, both at company level and, especially, on each PFC. For this purpose, it is necessary to continuously measure the non-productivity (losses and waste) and to transform them into costs (TCLW and TCCLW).

In order to achieve the comprehensive annual MCI targets setting, three levels of reconciliation are needed:

- Target price vs. annual target profit (level 1)
- Annual target profit vs. annual MCI targets (level 2)
- Annual MCI targets vs. annual MCI means (level 3)

Figure 3.4 presents the three levels of annual reconciliation of MCI targets for the four steps of approach to establishing the annual MCI targets and means (top-down approach: Market Driven for MCI and Profit Driven for MCI; bottom-up approach: Driven by MCI Targets Deployment and Driven by MCI Means Deployment). Reconciliation is performed routinely for each unit manufacturing cost structure (direct and indirect costs; transformation costs and material costs) and for each budget structure (fixed and variable costs), together with each type of losses and waste, implicitly with each structure of CLW and CCLW for each PFC and/or product at process and activity level, based on the market pressure (annual external profit and annual target profit) and on the annual MCI means.

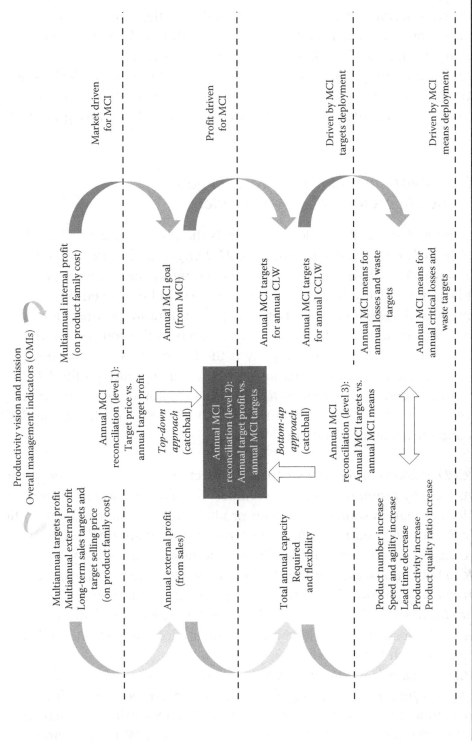

Figure 3.4   Annual reconciliation to set MCI targets for each PFC.

Reconciliation is accomplished with the aid of multiple simulations for a calendar year or whenever it is needed (e.g., twice a year, in August and in January), considering the significant changes during the year. The annual profit target must be met. Therefore, the annual MCI goal contribution to annual target profit can be adjusted even during the year.

In conclusion, the yearly reconciliation process of the annual MCI targets is the basis for uncovering hidden profits of annual profitability with the aid of continuous transformation of the manufacturing flow through annual MCI means.

3.1.4.2 Establishment of Annual MCI Targets and Means for Each Group Leader

An important element of the MCPD system is the involvement of all levels in both *planning items for setting annual MCI targets* and in *action items to set annual MCI means*. The exact understanding by the top management of MCPD and, implicitly, the level of MCI required to be reached on a multiannual and annual basis for both existing and future products for each PFC and the total involvement of lower levels in the company and other stakeholders beyond the company is the key to the success of the continuous transformation of a manufacturing company through the MCPD system. To this end, it is necessary to develop a consistent *cross-functional management* to implement the MCPD system and, implicitly, the annual MCI targets and means at all hierarchical levels:

1. *MCPD at board management level*: Connections between the *company productivity vision (CPV)* (sales volumes, the likely evolution of the prices of main resources and main products, and the multiannual profit plan) and the need *for multiannual internal profit* are established by increasing productivity and quality for the next three to five years.
2. *MCPD at senior managers level*: Connections between the *company productivity mission (CPM)* (the capacity need) and the *productivity core business goals (PCBG)* and the need for *multiannual internal profit* are set by increasing productivity and quality for the next one to three years for each PFC.
3. *MCPD at middle managers level*: Connections between *PCBG* and *company productivity strategy (CPS)* and the need for annual MCI goal (annual internal profit) are established by increasing productivity and quality for each PFC. For this purpose, annual MCI targets and

means and, implicitly, annual CLW targets and annual CCLW targets are developed for each PFC and for the entire company—using *MCI catchball process, MCICP*. In this way, each middle manager will be the owner of a certain number of kaizen projects (MCI means level 2) and kaikaku projects (MCI means level 3), which are considered strategic for that year to meet their *KPIs* target. Each KPI target is converged to the *annual MCI goal* and *multiannual target profit*. For example, the quality assurance manager, which is responsible for the *costs of scrap ratio losses* (KPI of the quality assurance department), as part of the current CCLW and/or CLW structure for a PFC, will have a *kaizen* project on the development of two Poka-Yoke devices, one for prevention and one for detection, for an activity in a process. It will be known from the very beginning as the level of CCLW and/or CLW necessary to be reduced throughout the entire manufacturing for PFC, by reducing the costs of scrap ratio losses and the annual MCI targets (e.g., \$75,000 in Table 1.3). In this way, all the resources needed for the *kaizen* project for Poka-Yoke will be allocated in time, through a rigorous timetable planning.

4. *MCPD at implementation teams level*: Each middle manager forms interdepartmental teams to run kaizen projects (MCI means level 2) and kaikaku projects (MCI means level 3) based on the *annual action plan* for each PFC. Each project will have clearly specified the start date and the completion dates and the expected benefits (especially reduced times, increased production capacity, reduced distances, reduced motions, reduced direct and indirect labor costs, and reduced stocks— taking into account the resources needed to be allocated).

5. *MCPD at the individual level*: Each member of a kaizen and/or kaikaku team knows exactly the *MCI means impact (MCImi)* (see Table 1.3), the required reduction of losses and waste in its area, the impact on KPIs of the targeted improvement, and the annual activities to be attended to achieve the annual MCI means (e.g., number of hours of training/ workshop/on-the-job training and the number of hours allocated for kaizen and/or kaikaku project; these hours will not be planned to contribute to the annual production).

Over time, each group leader can form interdepartmental committees to address specific departmental issues from the perspective of MCPD.

In fact, MCPD at implementation teams level is the most important phase of the MCPD system because it achieves the actual transformation of the current manufacturing flow from the CLW and CCLW perspective in order to meet customer satisfaction (productivity, cost/price, quality, delivery, environment, innovation), employee motivation (safety, health, and morals), and last but not least the need for profitability of the shareholders (more precisely, the multiannual and annual target profit through uncovered hidden reserves of manufacturing profitability with the help of annual MCI deployment and, implicitly, by reducing the CLW and, especially, CCLW through kaizen and kaikaku projects to increase productivity and quality). Achieving the annual MCI targets, implicitly annual CLW targets, and, especially, annual CCLW targets, meets all the required dimensions necessary for customer satisfaction, employee motivation, and consistent shareholders' need for profits.

In this context, the MCPD system becomes the quintessence of the harmonious transformation of manufacturing companies.

3.2 Preparation for the Implementation of the MCPD System

The need for the MCPD system is closely related to meeting the CPV and CPM, and especially achieving the multiannual target profit with the help of multiannual internal profit and annual MCI goal.

The customer satisfaction needs are perennial: product number increase, cost reduction, quality increase, and speed and agility increase. The cost reduction approach, that is, the manufacturing cost decrease, requires increased attention to all contamination related to the costs originating from the non-productivity, losses, and waste at the level of manufacturing flow and beyond. Manufacturing cost decrease can only aim the reduction in the costs behind losses and waste and not the costs perceived as having added value from the customers' perspective, eventually by lowering the quality level. The only way to satisfy customers through the expected price level and shareholders through the expected ROI is to continually reduce the cost by increasing productivity.

Therefore, the harmonious transformation of manufacturing companies through the MCPD system is continuously linked to the needs of the market, in order to ensure uncovering hidden reserves of profitability, namely, reducing and/or eliminating costs related to losses and waste.

3.2.1 *Decision to Implement MCPD and Prediction of Its Effects*

The success of the MCPD system depends on the *senior managers'* involvement degree, because any board of directors wants to maximize profits for their investments by minimizing inputs of resources and maximizing outputs of products and services.

Therefore, the person with the highest degree in the company should decide the company's transformation by emphasizing the pro-cost and pro-productivity thinking. To help this decision, strategic positioning or repositioning of the company is required from the point of view of the need for strategic cost reduction. This requires a clear understanding of the following:

■ Current and future policies and strategies and how they affect or will affect the company
■ The circumstances of the key current and future external and internal factors
■ The company's current performance compared to a reference value (benchmarking, best practice, or ideal state)

To help understand the specific processes of the MCPD system and its stake in supporting the company's long-term profit strategy, a pilot project for a family of products is being developed to analyze the concrete results achieved. Reducing the costs behind losses and waste will, in the first instance, aim at identifying the level of cost contamination of manufacturing costs with costs of losses and waste (usually at the beginning of the MCPD system implementation of at least 30%–35% of the manufacturing cost structure). For this purpose, the current costing and productivity monitoring system will be reviewed and the deviation from the ideal cost state will be determined to identify the uncovering hidden reserves of profitability and establish the main directions of manufacturing flow transformation.

Annual MCI targets will aim at meeting the annual MCI goal. The annual MCI goal is linked to both the current and future market needs for cost/price reduction and the need for annual manufacturing profit, as well as to the actual potential of meeting the annual MCI targets through annual MCI means at each PFC level.

The potential tangible effects of the MCPD system by reaching annual MCI targets are productivity increase, capacity increase, lead time decrease, and product quality ratio increase.

Therefore, ideally, the decision to implement the MCPD system originates from the person with the highest managerial level in the company. This written decision is made public through the internal communication system.

3.2.2 MCPD System Organization: Create Structures to Support MCI

Creating an organizational structure to support the MCPD system is necessary to exchange information on top-down and bottom-up costs with acceptable and timely accuracy to continuously support the company's objectives, mission, and vision on a continuous basis.

This structure will have the following main tasks:

■ Monitoring the progress of MCPD implementation in connection with the multiannual target profit, multiannual external profit, and multiannual internal profit
■ Participation in setting annual MCI goals, annual MCI targets, and annual MCI means to meet the PCBG
■ Monitoring the degree of reaching the annual MCI targets and means
■ Ensuring the accuracy of all data and information on losses and waste in all company processes
■ Ensuring the veracity of how to transform losses and waste into costs
■ Interpreting and reporting the current level of CLW and CCLW for each PFC, both in the annual MCI targets phase and especially in the phase of analyzing the degree of achievement of the annual MCI targets
■ Monthly verification of the annual MCI target fulfillment level for each PFC and, implicitly, the advancement degree of annual MCI means
■ Developing strategic directions to prevent the unfavorable evolution of the current CLW and, especially, the current CCLW associated with each PFC
■ Designing the resource budget required to meet the annual MCI means—including planning for people's participation in improvement projects (typically trying to assign at least 8% of the total time of people to participate in the achievement of the annual MCI means)

- Archiving the improvement projects and collaboration with other similar structures at group level
- Participation in master budget design, in designing the annual manufacturing improvement budget and analyzing the degree of reaching the annual MCI targets for each PFC, consistent with the annual MCI goal
- Ensuring communication with the MCPD consultant

Therefore, the main purpose of the MCPD organization is to provide all the information that is needed to remove any obstacle that may stand in the way of developing the pro-cost culture and, implicitly, the pro-productivity culture.

3.2.2.1 Commitment to MCPD

Following the decision to implement the MCPD system and setting the strategic expectations of continuous reduction of the manufacturing cost, we proceed to setting the annual MCI targets and means based on the annual MCI goals to achieve the annual target profit. For this purpose, the initial level of CLW is studied to determine the current state and to identify the necessary improvements. The organization and commitment to MCPD covers all factory departments: production, quality, maintenance, cost management, purchasing, material logistics, product development, production engineering, and human resources.

The formula for the continuous transformation of the manufacturing company through the MCPD system to achieve the annual MCI goal for each PFC and per overall company is

Problem awareness with CLW and CCLW

$$* \text{Annual MCI targets} * \text{Annual MCI means} = \text{Annual MCI goal} \quad (3.10)$$

The managers and leaders of these departments are the ones who continuously support the MCPD system through the following:

- Developing vision and mission statements related to the MCPD system at the department level; setting the main directions of the MCPD system for each department for the next three to five years
- Setting annual MCI targets and annual MCI means for each process of each PFC to be convergent to annual MCI goals; scientifically setting systematic (kaizen) and systemic (kaikaku) improvement directions

- Developing procedures for continuously capturing the current CLW and CCLW at the manufacturing process level of each PFC
- Ensuring all the necessary resources to meet the annual MCI means in time
- Providing procedures for capturing current results and for analyzing the possible differences between the annual MCI target and the current level of the unit manufacturing costs for each PFC and company-wide
- Developing training and workshop plans to support the continuous learning of all people in each department and beyond the company's borders

Managers and department leaders meet once a month to discuss and analyze the current state of the annual MCI targets and means, and once every three months to assess the possible proposals for new directions for meeting the annual MCI goal. Twice a year, any major rectifications of the annual MCI goal are made, depending on the changes inside and outside of the company.

3.2.2.2 Steering Committee of MCPD

The value of a manager lies in his/her ability to plan the actions and not to perform the necessary activities to carry out these actions. In order to plan the actions for the implementation of the annual MCI means targets, it is necessary to identify the direction and the means to achieve the set targets. In order to plan strategic cost reduction directions, it is first necessary to establish scientifically the annual MCI targets for achieving the annual MCI goal and then to trigger the annual MCI means at the most appropriate times. Otherwise, the annual MCI triggering without planning directions leads to the failure of people's confidence in the culture of continuous cost improvement and beyond.

Against this background, of the continued fight against the current level of unit manufacturing cost, through both frontal (CCLW, the acute cost problems) and encirclement (CLW, the chronic cost problems), it is necessary to establish a clear direction and identifying all the risks associated with the fulfillment of this direction that may stand in the way of meeting the annual MCI goal. For this purpose, it is necessary to designate an MCPD steering committee coordinated by a full-time task manager, which includes members at each PFC level as well as full-time tasks and other employees temporarily supporting certain MCPD activities and actions.

The MCPD steering committee meets daily and assesses the current status of the annual MCI targets and means planning for each PFC, evaluates daily cost management to identify the stage of problem solving for CLW and CCLW targets, and evaluates and re-plans the promotion of the MCPD system across the entire company. At the same time, every official or informal request from customers about the need for price reduction is captured on a daily basis (directly from customers or from the marketing department). The main goal is to capture any pressure on MCI as early as possible to have time to plan and meet the potential annual MCI targets in order to meet the annual MCI goal.

More specifically, the MCPD steering committee

- Assigns all KPIs related to *strategic key points on manufacturing processes* and, implicitly, *PCBG* to responsible managers, ensuring the balance between the required level of costs and quality and productivity for each PFC and/or product
- Assigns KPIs related to losses and waste and of the CLW and CCLW managers in each factory area responsible of each PFC—*kaizen* and *kaikaku* indicators (KKI) (the company is divided into physical areas)
- Develops training programs and provides support information on the MCPD system at all levels of the company (starting with top management)
- Develops training and workshop programs on methods, techniques, and tools for improving productivity and quality, creating a continuous learning organization
- Audits on a continuous basis how to determine CLW and CCLW for each PFC and for each area
- Participates in establishing the annual MCI targets and means and provides the necessary support
- Defines and continuously monitors the *desirable managerial behavior identity* at the MCPD system level (*management branding*) (Posteucă and Sakamoto, 2017, pp. 240–244), especially in the deployment of the annual MCI means
- Plans activities to support the motivation of MCPD implementation for all people in the company
- Sets the priorities for improving the MCI for each PFC and establishes the annual MCI means accordingly (annual MCI action plan planning)
- Evaluates the annual MCI targets (implicitly CLW targets and CCLW targets) and prepares and supports the monthly reports on the degree of fulfillment of annual MCI targets

- Prepares for approval the required annual MCI means plan (annual improvement budgets for each PFC)
- Evaluates the effectiveness and efficiency of training and workshop programs and annual MCI means
- Prepares the MCPD transformation plan together with the department managers to achieve the annual MCI goals, annual MCI targets, and annual MCI means for each process of each PFC (for the next three years; detailed development of the annual plan, from the current state of MCI to the required future status; quarterly action plan and monthly schedules)
- Supports periodical public presentations on MCPD achievements
- Draws up the MCPD manual for each PFC and other internal manuals on other methods, techniques, and tools for improvements
- Keeps in touch with the appointed MCPD consultant

In conclusion, the main role of the MCPD steering committee is to ensure the direction of the MCPD program, acting as a turn-key between the management team, the MCPD implementation teams, and every person in the organization and beyond, to achieve the multiannual and annual target profit, especially the internal profit and the annual MCI goal, by motivating, accelerating, and continuously pursuing all the necessary activities. Each member of the MCPD steering committee is an *agent of continuous pro-costs change.*

3.2.2.3 Forming an MCPD Implementation Teams

The creation of the implementation teams is required to carry out the annual MCI means targets. The cross-functional team is formed for annual *MCI means deployment* (kaizen and level 3—kaikaku) with the help of KKI. The size of a team varies between three and seven members and a single leader. Each member of the team is assigned a clear role. Usually, a representative of the cost management department is part of the team to verify the accuracy of the initial CLW and/or CCLW and to quantify the results. Each team should be as autonomous as possible and as objective as possible (based solely on actual data and facts gathered by it and managerially validated). Each team should be able to verify the performance achieved in meeting the MCI targets of the improvement project relative to the initial set level (knowing exactly what KPIs are targeted to be improved and the project's cash stake—reducing/eliminating the current level of CLW and/or CCLW). All members of the MCI team must be able to add value during the improvement project.

The manager of the MCPD steering committee meets weekly with the leaders of the implementation teams to identify the state of progress of MCI projects in order to make possible any resource unblocking and to provide any additional information needed. This meeting is also typically attended by managers in the area of the improvement project. At the same time, once a month, the state of the ongoing improvement projects, the results of the completed projects, the projects that follow, possibly the delayed projects together with the causes of their delay (projects started or not), and plans for the horizontal extension of the solutions successfully implemented (using one-point lesson—OPL) are displayed at the *MCPD information center.*

3.2.3 Internal and External Communication of the Purpose of the MCPD System

This initial communication of introducing the MCPD system (kickoff) is beneficial to provide implementation teams and external partners (major customers, major suppliers, and local government officials; members of the families of employees) as much information and clarification as necessary. The structure of the MCPD launch ceremony has the following structure:

■ Brief presentation of company history
■ Presentation of the current situation of OMIs and major external and internal trends
■ Presentation of the declaration of joining the MCPD
■ Reasons for choosing the MCPD system—presentation from all factory levels and functions
■ Tangible and intangible results expected from the MCPD system
■ Manner of choosing the MCPD steering committee members and their role
■ Expectations from all employees of the company from the MCPD perspective
■ Expectations from customers and main suppliers from the MCPD perspective
■ Presentation of the results of the MCPD pilot project and the presentation of the reasons for choosing the respective pilot project
■ Presentation of the resources that will be allocated for the transformation of the company through the MCPD system

■ Timetable for participating in MCPD training of all employees (starting with the management team)
■ Presentation of the master plan of the MCPD system
■ Presentation of the benchmark values and/or the ideal state of MCI in the pilot project
■ Presentation of the desirable behavioral identities expected from all the people in the company and beyond, especially in the case of managers
■ Free questions and answers session

Therefore, this internal and external communication has the role of shifting the rows to the fight with the continuous manufacturing cost reduction in order to ensure the level of competitiveness required by price and by profit. The stake of this communication is to immediately begin implementing the MCPD master plan to expand and amplify this emotional momentum of a new beginning in the history of the company.

References

Nakajima, S., 1988. *Introduction to TPM: Total Productive Maintenance.* Portland, OR: Productivity Press.

Ohno, T., 1978. *Toyota Production System—Aiming at an Off-Scale Management.* Tokyo, Japan: Diamond-Verlag.

Ohno, T., 1982. How the Toyota production system was created. *Japanese Economy,* 10(4), 83–101.

Ohno, T., 1988. *Toyota Production System: Beyond Large-Scale Production.* Cambridge, MA: Productivity Press.

Posteucă, A. and Sakamoto, S., 2017. *Manufacturing Cost Policy Deployment (MCPD) and Methods Design Concept (MDC): The Path to Competitiveness.* New York: Taylor & Francis.

Sekine, K., 2005. *One-Piece Flow: Cell Design for Transforming the Production Process.* Boca Raton, FL: Taylor & Francis.

Tapping, D., Luyster, T., and Shuker, T., 2002. *Value Stream Management: Eight Steps to Planning, Mapping, and Sustaining Lean Improvements.* New York: Taylor & Francis.

Womack, J. and Jones, D., 2011. *Seeing the Whole Value Stream,* 2nd edn. Cambridge, MA: Lean Enterprise Institute, Incorporated.

Chapter 4

MCPD Implementation: Designing, Building, and Full Development

This chapter focuses on the detailed presentation of the three phases and the seven steps of the MCPD system.

Section 4.1 presents the first phase of the MCPD system, *manufacturing cost policy analysis*, and the first two steps of MCPD, namely, setting annual MCI targets (step 1) and setting annual MCI means (step 2)—for each PFC by annual reconciliation (top-down and bottom-up for setting annual MCI goals and annual MCI targets and means). The basic goals are to achieve *total managerial commitment* and to reduce and/or eliminate *the resistance to change of all people in the company and beyond*.

Section 4.2 presents the second phase of the MCPD system, *manufacturing cost policy development*, and the next two steps of MCPD, namely, addresses the annual manufacturing improvement budgets development for existing and new products and the annual manufacturing cash improvement budget, in order to support the annual MCI targets and means (step 3) and the annual action plan for MCI means, including the annual individual plans for MCI (step 4). The underlying goals are to achieve *total involvement of all departments* to achieve continuous company transformation and to reduce and/or eliminate *reactive managerial behavior* to address operational challenges based on the annual action plan for MCI.

Section 4.3 presents the third phase of the MCPD system, *manufacturing cost policy management*, and the last three steps, namely, the engagement workforce to achieve the MCI targets through departmental organization for achieving MCI targets, through the development of an annual MCI training plan by running of the improvement activities and actions of the annual MCI means to meet annual MCI targets (step 5); the performance level of the current state of MCI against annual MCI targets is followed by developing the annual MCI performance management (step 6); and by daily MCI management (step 7), the tangible and intangible effects of annual MCI are monitored daily and the deviations of MCI targets and contextual managerial behaviors are solved. The basic goal is to reduce or eliminate *incorrect and incomplete improvement projects* implementation by continuously targeting these improvement projects through the need for unit manufacturing cost reduction imposed by the market (customers, competitors, and shareholders).

4.1 Manufacturing Cost Policy Analysis

The purpose of manufacturing cost policy analysis, the first phase of Manufacturing Cost Policy Deployment (MCPD), is to *define the problem* regarding the need for manufacturing cost improvement (MCI) and to *analyze the causes* that contribute to problem solving (*root cause of cost of losses and waste [CLW] and critical cost of losses and waste [CCLW]*), in fact to understand the necessary current and future conditions (*Step 1: Context and Purpose of MCI*) and to determine *the necessary countermeasures* by aligning the annual MCI means (*problem-solving techniques* [PST], *kaizen*, and *kaikaku*) to the annual MCI targets and further to the annual MCI goal (*Step 2: Costs Strategy into Action—MCPD Alignment for Setting the Annual MCI Targets and Means*).

4.1.1 Step 1: Context and Purpose of MCI

The pressure on the manufacturing costs level and on the remainder of the costs originates both from the market pricing practices, set under the healthy conditions of demand and supply manifestation, that is, cost principle of Toyota by Ohno "sale price − profit = costs" (Ohno, 1988, p. 8), and from the pressure coming from the shareholders by the expected percentage of *return on investment (ROI)* (revenue − cost of goods sold/cost of goods sold) throughout the duration of the investment and implicitly by attaining the multiannual target profit. *The only way to satisfy the two external entities of the company, the customers by the expected level of the price and the shareholders by the expected percentage of ROI, is to constantly reduce costs up to an expected level (acceptable for a certain moment) by increasing productivity*—more specifically, to continuously define, analyze, and identify the opportunities to fulfill the annual MCI goal, by continuously reducing CLW and CCLW, both for the existing and future products.

4.1.1.1 Strategic Key Points on Manufacturing Processes

Plan–do–check–act (PDCA) cycle, also called the Deming cycle, is the basic principle of the whole MCPD system. The first step (plan) aims at the identification of the operational improvements required to provide for the development of the annual MCI targets and means to fulfill the annual MCI goal and, implicitly, to support the company's productivity vision and mission statements. In order to meet the annual MCI targets, it is necessary to determine and analyze all the *strategic key points on manufacturing processes (SKPMP)*. From the point of view of the MCPD system, the analysis of *SKPMP* is required both for the *manufacturing and assembly type* of companies and for the *process type* ones, to fully understand the relationships between the MCPD system and the manufacturing flow.

This analysis is made for each product family cost (PFC) and is based on the elements of Formulas 3.4 (*ideal state of the manufacturing flow*) and 3.5 (*current state of the manufacturing flow*), that is, from total cycle time (TCT) and from work in progress (WIP) in setup time and WIP in transfer time, and on the elements of Formula 3.9 (*CCLW for manufacturing lead time of each PFC*)—cycle time of bottleneck vs. maximum takt time, process capacity, setup time, and transfer time.

In order to fulfill *their productivity vision* (multiannual target profit and sales volume) and *the productivity mission* (annual target profit, ensuring the manufacturing capacity with an acceptable level of quality and a CLW level that tends toward zero and a flexibility that might provide for the needs of the small volume of customers) and *productivity core business goals* (PCBGs)—*corporate goals* and the projections of *overall management indicators (OMIs)*, manufacturing companies develop *annual measurable objectives* intercorrelated for the next three years, such as income increase by X% annually, quality increase by Y% annually, reduction of CLW by Z% annually (or annual MCI goal), and the increase of annual flexibility by T%, such objectives that underlie the development of the *productivity master plan* (Posteucă and Sakamoto, 2017, pp. 21, 90–91).

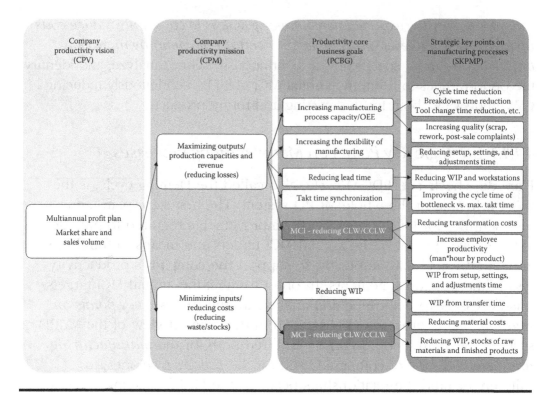

Figure 4.1   From productivity vision to SKPMP.

On such a background, Figure 4.1 presents an analysis of the SKPMP for each PFC, based on which the strategies, the annual objectives, the key performance indicators (KPIs), the kaizen activities, the kaikaku actions and the daily management will be developed later at all levels of the company.

Starting from the fact that the needs of focusing on one strategic key point or another, and implicitly the level of allocated resources, differ depending on the multiannual trend of sales, that is, if the *target profit* is monitored to be fulfilled predominantly from the maximization of outputs (if the sales trend is ascendent; *external profit*) or predominantly by the minimization of inputs (if the sales trend is descendent; *internal profit*), for the identification of SKPMP, it is a requirement to

■ Precisely define the contribution to profit of every manufacturing strategic key point
■ Determine the method of data collection, their intercorrelated verification, the trends evaluation, and their organization for the identification of the basic principles and phenomena for the following

period (revision of documents and information and interviews with managers and other key persons)

- Determine the real potential of contribution to the multiannual target profit of every manufacturing strategic key point
- Finally select and validate the contribution of every manufacturing strategic key point for the following period

The identification of the real potential of every SKPMP is done by means of the *catchball process*, more precisely by finalizing the decisions based on several rounds of meetings and discussions where all managers at all levels and all experts and all key persons in the company and beyond are involved.

Starting from the market and internal context and from actual records for PFC, more precisely from the current status of sales turnover and market share, manufacturing target profit, manufacturing cost, total lead time, and manufacturing lead time, by means of SKPMP, long-term business strategy and long-term cost strategy are defined for each PFC to take away the pressure on the costs corresponding to each PFC. Defining PFC is done by considering the following elements: (1) production numbers, (2) contribution to the multiannual and annual target profits for the annual MCI goal, (3) percentage of similar processes across the entire manufacturing flow, (4) level of the annual MCI targets, (5) products that have the similar potential to reduce CLW and meet the annual MCI targets, and (6) product life cycle is upward and/or fit into the initial volume and target profit (eventually).

The important points of SKPMP from the perspective of the annual MCI goal (reducing CLW/CCLW) are as follows:

- Determining the way of data collection concerning the actual level of losses and waste (see Section 3.1.3.2; overall equipment effectiveness [OEE] and OLE calculation) (Posteucă and Sakamoto, 2017, pp. 119–123)
- Intercorrelated verification of collected data (see Section 3.1.3.5; stratification analysis with Pareto diagram)
- Evaluating the trends of the collected data
- Organizing the collected data for the identification of the basic principles and phenomena for the following period

In order to achieve the evaluation of the data trends and their organization, it is necessary to identify exactly for each cost center the structure of unit manufacturing costs for every product and/or PFC (direct and indirect costs) and the expenditure structure (fixed and variable expenses) and, moreover, to determine the extent of affectation of these structures by the current level of CLW and CCLW identified at the process level of each PFC (the aim is to identify a minimum of 35% of the structure of unit manufacturing costs that are *infested* with CLW, *the chronic cost problems*, and CCLW, *the acute cost problems*). The allocation of costs for every cost center/process/work center/activity will contribute to the localization of costs, including of those with WIP. This analysis is made for every product/PFC, customer, process, work center, and activity.

Figure 4.2 presents an example of the evolution of the trend and structure of unit manufacturing costs (transformation costs and material costs) and of the current level of CLW for a product ("A") and the need for MCI (problem, non-value-added costs; see Formula 1.7), based on the evolution of the unit price and of the unit profit. As we can see, *unit manufacturing costs are contaminated/infested with CLW* (implicitly also with CCLW, which is a sum of CLWs), both at the level of transformation costs and at the level of material costs.

As we can see, at the beginning of year "N," CLW were measured and the annual MCI targets for *costs of losses for transformation cost* ($0.45/unit) and for *costs of waste/stocks–material costs* ($0.31/unit) were determined to attain the annual MCI goal at product level. Further on, at the level of the processes corresponding to product "A," the structure of CLW and CCLW was identified by the annual total CLW proposed to be reduced, based on *total losses and waste* (TLW) and on *total critical losses and waste* (TcLW). The total offer for MCI is much higher (TCLW and total critical costs of losses and waste [TCCLW]) than the annual MCI targets. This offer is obtained by the continuous measurement of non-productivity (TLW and TCLW) and its transformation into costs (TCLW and TCCLW). For product "A," the analysis of every unit cost structure, of transformation costs and material costs, and of each associated structure of CLW and CCLW is made. This annual analysis is called *annual quarantine analysis of CLW and CCLW* and aims at studying and thoroughly monitoring the principles, phenomena, and necessary parameters to set the annual MCI targets as accurately as possible and to choose the most efficient and effective annual MCI means, at the same time, to meet the annual MCI goal. The annual MCI means will aim at preventing the emergence and/or reemergence of the problem and meeting the set standard/pursuing the maintenance of the standard—PST, continuously improving by kaizen and improving by kaikaku. Similarly, the

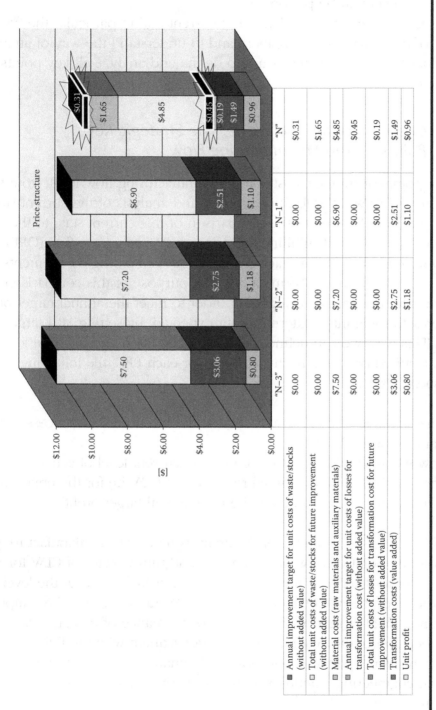

Figure 4.2   The evolution of the unit price vs. opportunities for annual improvement.

analyses for the current condition at PFC, customer, process, work center, and activity level will be performed.

Consequently, in order to define the current and in particular the future context of the business development and to understand the way of attaining the annual MCI goal, it is necessary to define and analyze all key points of strategic costs for each PFC.

4.1.1.2 Costs of Losses and Waste: The Strategic Key Points on the Manufacturing Flow

The key points of strategic costs on the manufacturing flow level, from the perspective of CLW, refer to the definition and analysis of the current state of CLW for each PFC and to the comparison of this current state to the ideal state under the terms of fulfilling the principles of *one-piece flow* (OPF) and *continuous flow* (CF). In fact, it is all about comparing the elements of Formula 3.5 with those of Formula 3.4. The purpose of this comparison is to define later the annual MCI goal in order to meet the annual target profit, together with the annual external profit expected, by setting MCI targets and means. This way, we can define the current condition and the total level of the identified problem. In this context, for each PFC, the following three main questions arise:

1. What is the level of CLW identified at the level of every process of the manufacturing flow?
2. How much should ideal CLW be for the current level of activity?
3. How much should the annual reduction of CLW be for the next three to five years in order to meet the multiannual target profit?

Table 4.1 presents an example of the analysis mode of the manufacturing flow from the point of view of the current and future states of CLW for each PFC in order to determine the future annual targets for CLW at the level of each process of the manufacturing flow (see Formulas 1.5 and 1.6), implicitly the future annual targets of MCI. For the performance of this analysis, the cooperation of the cost management department with the following departments is necessary: production, maintenance, quality assurance, industrial engineering, continuous improvement, internal logistics, and human resources.

The data and information on *current state* (column 1) are collected for each process by every department in charge. The main purpose

Table 4.1   CLW: Current Stage vs. Ideal Stage for PFC Processes

	Current State *(1)*	CLW of Current State *(2)*	Ideal State *(3)*	CLW of Ideal State *(4)*
Production numbers	100%	$X	125%	$X-30%
Waste: finished products stock	100%	$X	–98%	$X-92%
Process "n"				
Process 2				
Equipment losses (%, hours, and $)	44%	$X	8%	$X-35%
Human work losses (%, hours, and $)	25%	$X	1%	$X-35%
Material losses (% and $)	6%	$X	0%	$0
Energy losses (% and $)	4%	$X	0%	$0
Setup (minutes)	15	—	5	—
Total cycle time (TCT) (minutes)	*3 min*	—	*1 min*	—
Transfer time (seconds)	270	—	60	—
Number of workstations or operations	3	—	1	—
Takt time max/bottleneck (seconds)	22.5	—	5	—
Waste: WIP setup (WIP S) (minutes) or units	*30 min*	*$X*	*5 min*	*$X-83%*
Waste: WIP transfer (WIP T) (minutes) or units	*5 min*	*$X*	*1 min*	*$X-80%*
Process lead time (TCT + WIP S + WIP T) (minutes)	38 min	$X	7 min	$X-81%
Process 1				
Equipment losses: %	45%	$X	8%	$X-35%
Human work losses: %	26%	$X	1%	$X-35%
Material losses: %	5%	$X	0%	$0
Energy losses: %	3%	$X	0%	$0
Setup (minutes)	20	—	5	—
TCT (minutes)	*45 min*	—	*18 min*	—
Transfer time (seconds)	600	—	30	—
Number of workstations or operations	31	—	10	—
Takt time max/bottleneck (seconds)	1,975	—	1,200	—

(Continued)

Table 4.1 (*Continued*)   CLW: Current Stage vs. Ideal Stage for PFC Processes

	Current State (1)	CLW of Current State (2)	Ideal State (3)	CLW of Ideal State (4)
Waste: WIP setup (WIP S) (minutes) or units	20 min	$X	5 min	$X-60%
Waste: WIP transfer (WIP T) (minutes) or units	15 min	$X	5 min	$X-66%
Process lead time (TCT + WIP S + WIP T) (minutes)	80 min	$X	28 min	
Waste: raw materials stock	100%	$X	−92.30%	$X-90%
Waste: components stock	100%	$X	−91.70%	$X-90%
Waste: packaging stock	100%	$X	−88.90%	$X-86%

is to catch the current state of losses and waste with a high degree of accuracy as possible to be able to realize the main phenomena and restrictions for each PFC and to know the current structure of the *process lead time*.

The information corresponding to the *current state of CLW* (column 2) is provided by the cost management department that does the conversion between losses and waste into costs. These data and information on *current state of CLW* should be available at any time. It is important to obtain consistent data and information on the dynamics of the processes regarding CLW, such as

1. A gradual negative change of CLW (deterioration of CLW performances in time/gradual growth of CLW in time)
2. A sudden change of CLW (a radical change of CLW level; in particular, those that determine the sudden growth of CLW)
3. Periodical growth of CLW (with a relatively known repeatability; the main cause should be known, such as certain products, certain raw materials, certain shifts, for certain equipment, for certain periods of the year, etc.)
4. A consistency of CLW irrespective of the systematic and/or systemic improvements implemented
5. Improvements of CLW without having performed improvement projects in this respect (it will be verified in particular the accuracy of the collected data, but also the phenomena that generated the improvement of CLW)

Therefore, *the current state of CLW is in fact the current nonproductivity cost (lack of maximization of outputs, losses/lack of effectiveness and lack of minimization of inputs, waste/lack of efficiency).*

The information regarding *ideal state* (column 3) takes into consideration all the systematic and in particular systemic improvements that might be implemented in order to eliminate all and any restrictions at the level of each process of the manufacturing flow, in order to attain the OPF and CF states. This state is achieved, or better imagined, by the technical staff and by specialists from the company's headquarters, considering also the sales forecast and implicitly the current and/or necessary capacity analyses.

The information on *ideal state of CLW* (column 4) is that which considers the challenge to attain *zero costs of losses and waste (ZCLW) state.* The concept of zero is approached as being zero over the current admissible maximum, such as in Table 4.1, for process 2, zero over the five minutes to setup and further, lower the setup time as much as possible to zero minutes. In this case, it is considered that the five minutes to set up cannot be eliminated completely because at that time no technical solution (equipment) is known that does not require setup activities in that process. The purpose of the ideal state of CLW is to continuously provide the ideal perspective on the continuous improvement activities (kaizen; planned based on a Gantt chart) and on the systemic improvement activities (kaikaku; planned based on a Gantt chart) and on the cost/profit analyses concerning the choice of the best annual strategic directions to meet the annual MCI targets. For example, the ideal state of CLW should continuously provide for a minimum of 35% of CLW for the following five years, which might reduce by systematic and systemic improvements of productivity. Further, based on this percentage of the ideal state of CLW, the most feasible annual improvements are analyzed (annual MCI means). At the beginning of the implementation of the MCPD system, this annual percentage of CLW reduction may be of the minimum 6%. Then, in time, against the background of reducing the gap between the current state and ideal state of CLW, this percentage may decrease. This percentage is an input information to determine the annual target profit and the annual external profit (for the annual development of the master budget). It is important to know at any time the percentage of CLW at the level of each process in order to determine the approachable percentage of CLW for each product/PFC. It is worth noticing that an increased attention must be paid to allocating the costs at the level of each process and to allocating as much as possible direct costs and especially manufacturing overhead costs at the process level and then at the product/PFC level.

Therefore, by *strategic key points on the manufacturing flow* from the perspective of CLW, it is understood to identify the opportunities of reducing CLW at the level of every process for each PFC by comparing the current state with the ideal state. In order to determine the current and the ideal state of CLW, it is necessary to monitor and analyze in detail the losses and waste for all the activities of each and every process. The basic reasoning is: *the more CLW and implicitly losses and waste decrease, the more the productivity level increases and the level of unit manufacturing costs decreases, both against the maximization of outputs/minimization of losses/maximization of effectiveness (the increase of the current capacity use and implicitly the decrease mainly of the manufacturing overhead costs), and against the minimization of inputs/minimization of waste/maximization of efficiency (minimizing excessive input quantities and implicitly the decrease mainly of manufacturing direct costs—raw material costs and direct and indirect labor costs).* From the perspective of MCI, it is important to know at the time of determining the annual target profit and implicitly the annual MCI goal, which is the prevailing source of meeting the annual MCI targets and implicitly the annual CLW targets, that is, predominantly from the reduction of *manufacturing overhead costs (in the context of an increase of sales volume, to support the increase of annual external profit; the price being relatively fixed) or predominantly from the reduction of raw material costs and direct and indirect labor costs (in a context of stagnation or sales reduction, in order to complete the reduction of annual external profit). This is the approach of key points of strategic costs on the manufacturing flow.*

The basic challenge is to continuously target the CLW improvement by choosing the most feasible activities and improvement actions (annual MCI means) and their framing in the required time limit (until the level of unit manufacturing costs must drop, which should ensure a certain level of acceptable competitiveness, through price and internal profit, for the following period).

4.1.1.3 Costs of Losses and Waste: The Strategic Key Points on Process and Activity

As we can see in Figure 4.1, in section PCBG, MCI and implicitly the reduction of CLW and CCLW are to be found both in the case of fulfilling the multiannual target profit by outputs maximization and by inputs minimization. The purpose of the analysis of the strategic key points on the process and activity level of CLW is to identify the opportunities of reducing the unit manufacturing costs, both for transformation costs and for material costs. Both the existing and future products that will experience the same process will benefit from such opportunities.

Therefore, in order to identify such opportunities, we should start from the elements of Formula 3.6 (CLW structure for every process) presented previously in Section 3.1.3.3. Figure 4.3 presents an example of calculation and analysis of the CLW structure for a manufacturing process of PFC, with the activities generating associated CLW, for four similar products, for the average of the last six months. In this example, the equipment scheduled downtime is considered to have no influence that may be approached for the time being to reduce CLW.

Notes:

1. Equipment working hours = 14,400 minutes; value-adding operating time (VAOT) = 11,872.9 minutes; equipment losses = 1,147 minutes (19.1 hours/month); the calculation method for OEE was similar to Table 3.1.
2. Variable transformation costs (VTC) + fixed transformation costs (FTC) = $13.71/minute. Material costs (MC) = $239.49/product.
3. Manufacturing unit cost = VTC + FTC + MC = $253.2/unit.
4. Current admissible losses (CAL) = 780 minutes; current inadmissible losses (CIL) that create waste (stocks) = 367 minutes (equipment losses – CAL; 1,147 minutes – 780 minutes = 367 minutes).
5. Missed opportunities for producing pieces (finished pieces for the equipment) = 250.66 pieces (standard cycle time * CIL that create waste; 0.683 minute/pieces * 367 minutes = 250.66 pieces).
6. The equivalent in $ of the stocks brought "too early" for the 367 minutes (plus other costs):
   a. Raw material = $57,250.08 (350 minutes from 367 minutes; 0.683 minute = 1 piece; 350 minutes * 0.683 minute/piece = 239.05 pieces [for which the equivalent of raw material from the warehouse is calculated]; number of pieces * average unit cost of raw material stock = 239.05 pieces * $239.49/product = $57,250.08).
   b. Components = $59.76 (350 minutes from 367 minutes; 0.683 minute = 1 piece; 350 minutes * 0.683 minute/piece = 239.05 pieces [for which the equivalent of components from the warehouse is calculated]; number of pieces * average unit cost of components stock = 239.05 pieces * $0.25/piece = $59.76).

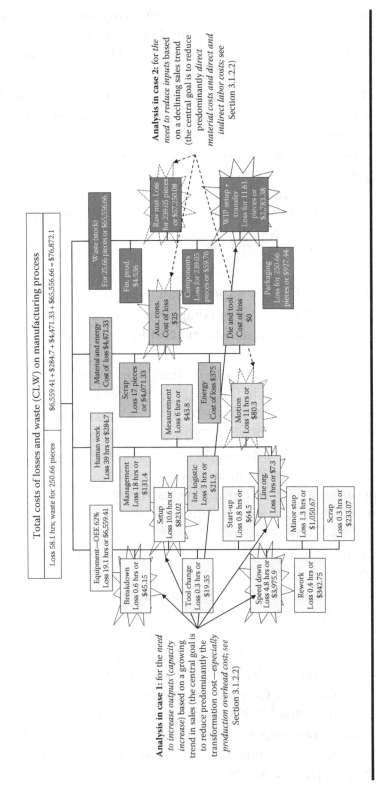

Figure 4.3 Detailed analysis of the CLW structure for a manufacturing process of a PFC (cost center).

c. WIP from setup and transfer = $2,783.38 (17 minutes; 0.683 minute = 1 piece; 17 minutes * 0.683 minute/piece = 11.61 pieces [for which the equivalent of WIP is calculated]; number of pieces * (average unit cost of raw material stock + average unit cost of components stock) = 11.61 pieces * ($239.49 + $0.25) = $2,783.38).

d. Packaging = $927.44 (missed opportunities for producing pieces [250.66 pieces] * packaging unit cost; 250.66 pieces * $3.7/pieces = $927.44).

e. Finished products stock = $4,536 (the cost of a cubic meter of finished products stock for 1 hour * number of hours of finished goods stock; $27/hour * 168 hours = $4,536).

7. Human work losses: 1 hour for direct and indirect labor at P2 = $7.3/hour.

a. Management loss: man * hours/month * 1 hour for direct and indirect labor = 18 hours * $7.3/hour = $131.4.

b. Motion loss: man * hours/month * 1 hour for direct and indirect labor = 11 hours * $7.3/hour = $80.3.

c. Internal logistic loss: man * hours/month * 1 hour for direct and indirect labor = 3 hours * $7.3/hour = $21.9.

d. Measurement loss: man * hours/month * 1 hour for direct and indirect labor = 6 hours * $7.3/hour = $43.8.

e. Line organization and balancing loss: man * hours/month * 1 hour for direct and indirect labor = 1 hour * $7.3/hour = $7.3.

8. Cost of loss for material and energy:

f. Scrap (material) loss = 17 pieces × MC (P1 + P2) = 17 pieces/month * ($239.49/product) = $4,071.33/month.

g. Auxiliary consumables loss = actual cost (P2) − theoretical cost (P2) = $25.

h. Energy loss: actual cost (P2) − theoretical cost (P2) = $375.

i. Die and tool loss: The number of units produced between the two die and jig changes for P2 = $0.

CAL means the current acceptable level of losses for the equipment for a given time, as in the manufacturing environments, it is the equipment that determines the level of value-added activities in the process (VAOT). At the same time, the annual MCI targets are set for all waste due to CAL overrun. CAL is the basis of the annual budgetary procedure of the external profit expected and at the basis of production planning (determining the probable capacity of the equipment). When setting the annual MCI targets, it shall be taken into account that not all CLW related to each manufacturing process may be identified, even in their ideal state. There are always unforeseen events that affect the level of CLW, such as a breakdown caused by an operator.

For a start, the equipment considered critical (strategic or bottleneck) shall be chosen to determine CLW. Nevertheless, CLW may be determined for all manufacturing processes for each PFC, even if the effort is sometimes considerable and a careful analysis of the costs involved and the profits obtained is required.

As can be seen in Figure 4.3, the analysis of CLW on whose basis the annual MCI targets shall be set at the level of a process of a PFC is influenced by the two hypostases of the sales (growth or decline).

So, the analyses are as follows:

■ *Analysis in case 1, for the need to increase outputs (capacity increase)* based on a growing trend in sales: The central goal is to reduce predominantly the *transformation cost, especially production overhead cost* (see Section 3.1.2.2), in order to contribute to the annual MCI goal. The opportunities identified for the annual MCI targets at the process level in order to support the achievement of the annual MCI goal are for the cost of breakdown losses ($45.15), cost of setup losses ($823.02), cost of speed down losses ($3,975.9), cost of line organization and balancing ($7.3), and costs of WIP from setup and transfer ($2,783.38). The collateral effects are also at the level of increasing the volumes requested by the customers in the context of obtaining a competitive price, in particular in the case of manufacturing overhead cost decrease. Usually, the more VAOT grows, in the context of the increase of OEE percentage, and implicitly of the volume of parts processed with the quality requested by the customer, the more the unit transformation cost decreases, especially unit manufacturing overhead cost, and CLW decreases. Nevertheless, by a proactive and preventive approach, the MCPD approach is to set beforehand the annual MCI targets and means in order to direct all the required improvements and resources, *neither more nor less than it is needed at that time,* to achieve the annual and multiannual target profits by means of annual and multiannual internal profits by total and continuous involvement of all the people in the company and beyond.
■ *Analysis in case 2, for the need to reduce inputs* based on a declining sales trend: The central goal is to reduce predominantly *direct material*

costs and direct and indirect labor costs (see Section 3.1.2.2), in order
to contribute to the annual MCI goal. The opportunities identified
for the annual MCI target at the process level in order to support
the achievement of the annual MCI goal are for the cost of auxiliary
consumable losses ($25), cost of raw material losses ($57,250.08), and
cost of motion losses ($80.3). The approach of these opportunities aims
at limiting the negative effects of overcapacity (especially of equipment)
in the context of reducing the sales volumes and implicitly of the
production volumes.

So, depending on the probable evolution of the sales in the following
period, the outputs increase and inputs decrease will be predominantly
emphasized. The determination of the percentages for the annual MCI
targets will take into account the probable evolution of sales in the
following period. The main purpose is to ensure an acceptable level of the
annual target profit, regardless if the sales are decreasing or increasing,
and to direct accordingly all the resources required to achieve the annual
MCI means (especially kaizen and kaikaku), in a planned way (both
annually and multiannually). Nevertheless, for a certain period of time at
the beginning of the implementation of the MCPD system, for example,
for a year, irrespective of the potential upward or downward evolution
of sales, the annual MCI means (e.g., kaizen for setup) will approach all
the opportunities considered feasible at that time, and the difference is
given only by the percentage of the annual MCI targets obtained from the
analysis in case 1 or *analysis in case 2.* For example, for a manufacturing
process, for the following 12 months when a growth of sales by 15% is
predicted and, implicitly, of a need to unlock current capacity, a reduction
percentage of 18% of annual CLW corresponding to transformation cost
may be set (especially manufacturing overhead cost), while the reduction
percentage of annual CLW directly corresponding to material costs
should be at 2.5%, considering the real current possibilities, especially the
supplier prices.

The choice of opportunities for the annual MCI targets mentioned
previously was made based on the analysis of the annual MCI means
(levels 2 and 3), at the same time or after the choice of these opportunities,
by the annual MCI catchball process (MCICP). Nevertheless, the annual
MCI targets may be set for all the opportunities identified at the level of
each process. From the perspective of unit manufacturing costs, the basic
reasoning is: if the *costs of current inadmissible losses and waste* from the

process P1 by $X will be reduced (i.e., CLW that may be approached), implementing the identifiable solutions by means of level 2 (a kaizen project), then the impact of the reduction on the total unit manufacturing costs will be at $Y and the cost of implementation will be at $Z. This way, the total contribution to the reduction of unit manufacturing costs for every manufacturing process of a PFC is set (within CLW are also included the CCLW—see Formula 1.6).

The choice of opportunities for the annual MCI targets (*CLW problems statement*) is done based on *root cause analysis* (RCA) for the identification of the roots of the sources of contamination/infestation of unit manufacturing costs with CLW, both at the level of transformation costs and at the level of material costs. Figure 4.4 presents the relationships between the effect and causes of CLW at the level of manufacturing process and of associated activities.

As can be seen, several types of causes are analyzed:

- *Direct cause*: Are those causes that give answers to the first question *Why*? The question is: *Why are there CLWs in this process*? Because *costs of equipment losses, costs of human work losses, costs of material and energy losses*, and *costs of waste* (stock) are present.
- *Contributing cause*: Answers are provided for the next level of *Why*? questions. The questions are of the following type: *Why are there costs of equipment losses*? Because cost of breakdown losses, cost of setup/adjustment losses, cost of cutting-blade losses, cost of start-up losses, cost of minor stoppage/idling losses, cost of speed loss, cost of defects, and cost of rework losses are present. Similarly, questions are asked for *costs of human work losses, costs of material and energy losses*, and *costs of waste (stock)*.
- *Root cause for visible effects on the process*: Provide the answer for losses and waste and for associated costs (to the question *Why*?). The questions are of the following type: *Why are there costs of breakdown losses*? Because (1) breakdown losses and (2) assignment of costs associated with breakdown minutes are present. *In this case, RCA is done vertically with the help of the "Why?" questions.*
- *Root cause for effects on the manufacturing flow*: Provide the answer to identify the *RCA* along the manufacturing flow for the identification of the sources of contamination/infestation of the unit manufacturing costs with CLW, which may have effects on the level of a PFC or another. *In this case, RCA is done horizontally with the help of "Why?" questions.*

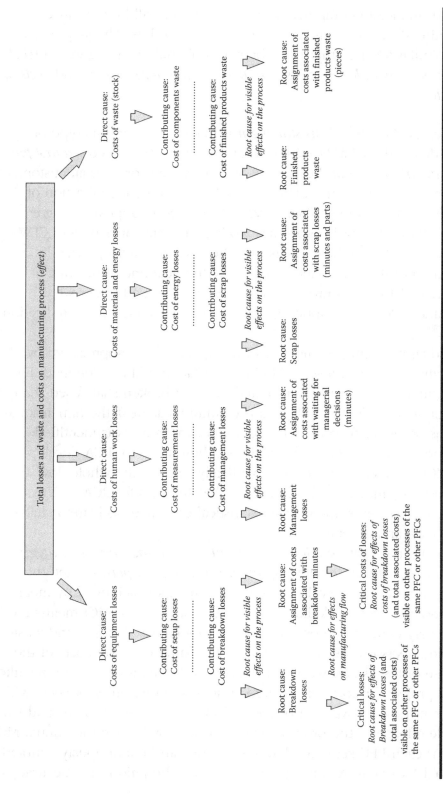

Figure 4.4 Mapping CLW causes for manufacturing process, activities, and manufacturing flow.

For example, an equipment breakdown event will be the cause of certain effects along the entire manufacturing flow. The judgments are similar for the other three branches of analysis in Figure 4.4 (*costs of human work losses, costs of material and energy losses,* and *costs of waste* [stock]).

Therefore, the *CLW problems statement* is made based on this type of RCA. To define the *CLW problems statement 5W + 2H* is used—*a tool for problem definition*: *Who?* the identity of external and/or internal customers who "complain" against the CLW level; *What?* accurate identification of CLW; *When?* axis of time (When have CLW started to appear?); *Where?* the localization of CLW in processes/work centers/activities (where CLW is produced, and not where it was identified; at root cause); *Why?* the identification of the explanations regarding the deadline of meeting CLW targets (Why is it a problem now?); *How?* How and under what circumstances does CLW appear?; *How many?* magnitude or quantification of CLW.

Individual problem statement of CLW targets will be approached by MCI means, when the root cause of CLW manifestation will be identified. For example, in a kaizen project for breakdown of a broken shaft/used shaft, as a result of RCA to identify MCI means (countermeasures), the dimensional deviations of the shaft were analyzed and a root cause was identified: nonconforming execution of the shaft. So, these types of RCA develop within MCI means for a process, or vertically, and for the whole flow, or horizontally (by means of the consecrated techniques for RCA: 5 Whys, fishbone analysis, is/is not analysis, statistical data analysis [Pareto charts, ANOVA, etc.], and others).

Therefore, the *CLW problems statement aims at the annual strategic steering of MCI means for meeting the annual MCI targets.* In fact, the targeting of all the nonproductivity improvements is achieved by means of CLW. Some kaizen or kaikaku improvement projects (MCI means levels 2 and 3) may concentrate on costs of the root cause of losses and waste for visible effects on the process and, at the same time, on costs of the root cause of losses and waste for effects on the manufacturing flow, like a kaizen project for costs of breakdown losses, while other projects will aim only at the improvement of costs of the root cause of losses and waste for visible effects on the process. These projects will have the annual MCI targets for each PFC set beforehand accurately and continuously coupled to the annual MCI goal. A constant attention must be paid to the accuracy of

CLW and to the market information for the steering to be relevant and to assure the expected results.

So, by summing up these scientifically identified opportunities (not based on personal opinions, more or less objective) and considered approachable to a certain percentage, by all the persons who can intervene on them at the level of each process/area, the level of annual reduction of CLW and CCLW will be determined to support the competitiveness through price/manufacturing cost and through profit (annual MCI goal) of each PFC.

4.1.1.4 Critical Costs of Losses and Waste: The Key Points of Strategic Costs on the Manufacturing Flow

The evolution in dynamics of a process, implicitly of associated CLW, is influenced by the evolution in dynamics of other processes. In order to set the annual MCI targets, based on a RCA for the whole manufacturing flow of a PFC, it is necessary to analyze also the level of evolution in the dynamics of annual CLW of every process and implicitly of CLW from other processes that influence other processes (downstream, upstream, and/or parallel). So, the influence in time of the behavior of losses and waste in a manufacturing process on other losses and waste in another process or on other processes of a PFC or on another PFC determines modifications on the levels of CLW. In this context, CCLW means all the costs generated by losses and waste in a process, both within the process in question and especially on the other processes along the whole manufacturing flow of a PFC or another and beyond (on the supply chain).

Therefore, the annual MCI targets will aim first at approaching the CCLW (the acute cost problems; root cause for visible effects on the manufacturing flow, losses and waste, and associated costs along the manufacturing flow— RCA approach horizontally) and then the CLW (the chronic cost problems; root cause for visible effects on the process, losses and waste, and CLW on process—RCA approach vertically).

Figure 4.5 is an example of identification of the sources of losses and waste likely to be root causes of a part of CLW in process 2. It presents the query mode of the manufacturing flow in order to discover the root cause for visible effects on the manufacturing flow by means of the 5 Whys technique. As can be seen, the causes/sources of the contamination outbreak of a unit manufacturing costs with CLW in process 2 may be multiple. Thus, to identify the root cause, the manufacturing area may be overcome, and a holistic approach is required to locate and determine the process-level CLW.

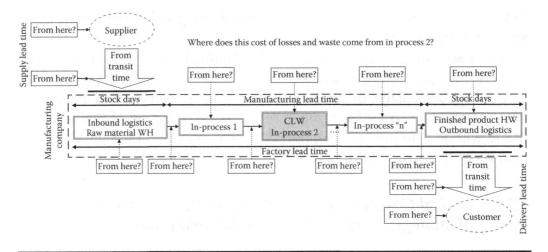

Figure 4.5   The holistic approach to multiple causal relationships in determining process-level CLW.

In this context, the unit manufacturing costs in process 2 are composed of the following:

1. Transformation costs and material costs (costs charged with added value in the present, even if they are bearers of certain CLW structures at the cycle time of the equipment and operators level; their approach is done by means level 2—kaizen or by *methods design concept*, MDC) (Posteucă and Sakamoto, 2017, pp. 327–354)
2. Transformation costs and material costs (CLW associated to the activities of process 2)
3. Transformation costs and material costs (CLW that were discovered in process 2 but have as root cause the activities of other processes, manufacturing process or not, coming from processes considered to generate CCLW)

All the three previously mentioned cost structures need to be standardized in order to be stabilized and to be predictable in order to comply with the annual MCI targets. In this context, of manufacturing costs standardization and stabilization to meet the annual MCI targets, an inside-out approach is needed:

1. *Shop floor process* (production and maintenance; men and machine; especially for direct labor costs and for manufacturing overhead costs)
2. *Manufacturing process* (quality assurance, internal logistics, and production control; especially for material costs, indirect labor costs, and manufacturing overhead costs)

3. *Warehouses process* (raw materials warehouses, components warehouses, parts warehouses, finished product warehouses)
4. *Supply chain process* (inbound logistics, outbound logistics, sourcing, customer distribution)

From the perspective of the MCPD system, CCLW deals with the first three types of the above processes. Nevertheless, a question then rises: *which are the main types of losses and waste considered to be the root cause for visible effects on the manufacturing flow?*

The phenomena that determine the outbreak of such losses and waste and impact on all the processes and on the entire production system are of two types:

1. *The lack of standard or significant variations of the existing standards*: Examples are equipment setup and adjustment times; planned maintenance times; cleaning/lubrication/inspection time; material handling time (man-hours spent to transport raw materials, components, scrap, rework, auxiliary materials); (re)start the equipment time; standard cycle time; counting parts/components, minimum and maximum for WIP, man-hours needed to adjust the machine, and including quality control checks; etc.
2. *Unplanned equipment/process failures*: Examples are equipment breakdown time, time of employees' sickness, line organization—failure to automate or poor automation

Going back to Formula 3.9 (CCLW for the manufacturing lead time of each PFC), for the processes standardization and stability along the manufacturing flow in view of the reduction of the variations of CCLW, it is necessary to reduce and/or eliminate the impact of CLW variations (variation of standard cycle time of bottleneck, variation of standard process capacity, variation of standard setup time, and variation of standard transfer time) on all the manufacturing processes and beyond them in view of meeting the annual MCI targets. CLW variation (*bottleneck, process capacity, setup time*, and *transfer time*) in a certain process stands for the causes of the effects on CLW occurred in the chain, on the upstream processes, and in the downstream manufacturing flow and beyond the manufacturing flow, for a PFC.

In order to choose a CCLW that needs improvements, thresholds of significance are set for every CCLW from the level of every PFC.

For example, in the case of five processes (five equipment/cost centers) that were designed to work as OPF, the moment a breakdown event occurs in process 3 (equipment 3), processes/equipment 1 and 2 stop or slow down and processes/equipment 4 and 5 do the same.

The determination of CCLW is done by summing up CLW corresponding to costs of breakdown losses in process 3 (*variation of standard process capacity—equipment failures*), because the following costs corresponding to equipment breakdown losses are activated:

- *Scrap cost*, incomplete parts or parts broken in process 3
- *Direct labor costs*, operators in process 3 who have no tasks, even if they can be moved to other posts
- *Spare parts cost*, parts replacement per equipment
- *Indirect labor costs*
- *Possible costs of external maintenance services*
- *Auxiliary materials cost*
- *Utility costs*
- *Depreciation costs*, with equipment in process 3
- *Costs of WIP from transfer* (between process 3 and process 4)

At the same time, for a complete determination of CCLW, the following costs corresponding to CLW are summed up (which were spread by the equipment breakdown losses from process 3 further in processes 1, 2, 4, and 5):

- *Direct labor costs*, operators in processes 1, 2, 4, and 5, who have no tasks, even if they can be moved to other posts
- *Indirect labor costs*
- *Auxiliary materials cost*
- *Utility costs*
- *Depreciation costs*, with equipment in processes 1, 2, 4, and 5
- *Costs of WIP from transfer between all processes (costs of waste)*
- *Modification/increase of costs of raw materials stock days (costs of waste)*
- *Modification/increase of costs of components stock days (costs of waste)*
- *Modification/increase of costs of packaging stock days (costs of waste)*
- *Modification/increase of costs of finished product stock days (costs of waste)*

For a complete determination of CCLW (which were spread by equipment breakdown losses in process 3), the costs corresponding to CLW beyond *factory lead time* may be summed up, on the whole *total lead time*, more precisely, on *material lead time* and *delivery lead time*.

Considering the monthly average time of breakdown losses in process 3, for example, of 563 minutes (frequency and average duration of an event) and the impact of this time on all CLW in all manufacturing processes and beyond, CCLW is quantified to an amount of $198,457/year. As the predominant causes of breakdown losses are mechanical failures, the planning of a kaizen project is imperative for breakdown/mechanical failures (MCI means level 2) for process 3 in view of reducing the amount of $198,457/year. This value is only a CCLW. The challenge is to identify all the relevant CCLW in order to slowly attain *CLW ideal state and to attain the annual MCI goal by the annual MCI targets*. All these CCLW are the strategic cost key points on the manufacturing flow and will make the object of the annual MCI means (in particular levels 2 and 3). The target of reducing the amount of $198,457/year will be set depending on the need for the annual MCI goal for PFC (see Formulas 1.5 and 1.6).

So, in accordance with Formula 3.9, CCLW has three levels of analysis:

■ *CCLW level 1 analysis*: Comparing the difference between the *standard values of CLW*, acceptable values for a certain given time, for bottleneck, process capacity, setup time, and transfer time and *their current state* at a certain time, and setting the directions of elimination of these variations (annual MCI means deployment level 1)

■ *CCLW level 2 analysis*: Comparing the difference between the *annual CLW targets* for bottleneck, process capacity, setup time, and transfer time and *their current state* at a certain time, and setting the directions of standardization (SOP) of these variations (annual MCI means deployment level 2)

■ *CCLW level 3 analysis*: Comparing the difference between *the ideal values of CLW* for bottleneck, process capacity, setup time, and transfer time and *their current state* at a certain time, and setting the directions of standardization (SOP) of these variations (annual MCI means deployment level 2 and 3). In the ideal state of manufacturing lead time, it is deemed that there is no CLW because the *OPF* state is fulfilled, the *CF principle* is observed, and the *zero cost of losses and waste* target is attained (see Formula 3.4).

The basic logic of the MCPD system is to set the annual MCI means for the root causes of CLW and especially of CCLW and not to set the annual

MCI means for the effects that are seen at the level of CLW in a process or another. The approach of the three levels of CCLW stands for the opportunities of *uncovering hidden reserves of profitability* for a process, for each PFC, and for all the company (see Formulas 1.8 through 1.10). So, in the context of continuous reduction of CLW and CCLW, even if in time their structure is continuously changing and they catch the real state of the manufacturing flow for a certain period for a PFC, unit manufacturing costs tend to drop sharply by implementing solutions identified by the annual MCI means. Based on these three levels of reference, standard level, target level, and ideal level, plans and scenarios are developed for the future states of MCI and implicitly of CLW and CCLW, which might assure the level of unit manufacturing profit required and meeting the annual MCI goal.

4.1.1.5 Analysis of the Manufacturing Flow to Establish Productivity and MCI Strategies

In order to set productivity and MCI strategies, a detailed analysis of manufacturing processes is required. This analysis starts from the awareness of the endless destination of MCI, from SKPMP, and from actual manufacturing and costs records corresponding to each PFC. Based on these, we can define the main phenomena perceived at the level of the manufacturing flow and we can determine the main working hypotheses in view of setting the annual MCI targets and means.

Therefore, in order to set the annual MCI targets and means based on productivity and MCI strategies and implicitly in order to fulfill the need for cost/price reduction and for target profit for the following one to three years, the following is required:

■ *Full comprehension of logical connections of current manufacturing and costs records* (of OMI with KPIs for costs and for CLW; see Table 1.3): a holistic thinking of the whole manufacturing flow of each PFC is needed that should aim at the following strategic elements: (1) reduction of manufacturing lead time, (2) synchronizing all the processes at takt time, (3) reduction of WIP stocks, (4) developing the OPF state in as many manufacturing flow areas as possible, if not entirely, and (5) heading toward the *zero costs of losses and waste* state. The current stage of performances of the current year and the extent of attaining the targets must be evaluated, in order to insulate every problem from the past. Every problem needs the definition of the root causes and the confirmation of the multiple effects of these root causes. Then, by means of Pareto analysis, priorities are set in choosing the annual MCI means to meet the level of the annual MCI target expected and further to fulfill the annual MCI goal.

■ *Detailed analysis of the SKPMP* (see Figure 4.1) for the identification of process weaknesses (see Table 1.3), in a holistic approach of process-level CLW (see Figure 4.3). For example, following this analysis, in which the whole department takes part, for *increasing manufacturing process capacity/OEE* (PCBG) by *cycle time reduction, breakdown time reduction, tool change time reduction, etc.*—strategic key points on manufacturing processes (SKPMP)—the following strategic directions are identified for the later determination of the annual MCI targets and means:
   – Planning work measurements at regular intervals and the elimination of non-value-added work that is not in the work procedures
   – Continuous measurement of OEE for selected equipment
   – Determining the resources needed to eliminate the breakdown (e.g., improvement projects using PM analysis)
   – Developing standards for new and existing equipment for tool change time, setup and adjustment time, start-up time, etc.
   – Designing new products to ensure optimal capacity loading (easy to produce)
   – Purchase/design of new equipment that is effective and efficient
   – Developing training plans for each operator so that each person performs his/her regular tasks and additional tasks
■ *Awareness and prioritization of the destinations for attaining the annual MCI targets* (see Figure 2.7). The order of priorities is chosen from among the following: product number increase (MCI means), quality increase (MCI means), and speed and agility increase (MCI means).

From the perspective of the MCPD system, the purpose of setting the *productivity and MCI strategies* is to support *company policy deployment* (increase the number of products, cost reduction, increasing product quality) at the level of

■ KPIs target of all plant components of the group
■ KPIs target for each PFC of the plant
■ KPIs target for each process level of CLW and CCLW
■ KPIs target for each activity level of CLW and CCLW

So, in order to meet the *company policy deployment*, targets synchronization is needed, bottom-up and top-down, for the following plant components in the group: department, section, division, work center, equipment, and people. Meetings will be organized at the managerial level for them. These meetings have a well-defined frequency and duration.

Productivity strategies for each PFC are set on three levels: level 1, the basic productivity strategies; level 2, the departmental productivity strategies; and level 3, the product family productivity strategy. An example of productivity strategy on three levels for the earlier example for *increasing manufacturing process capacity/OEE* (PCBG) is as follows:

- ■ *Level 1, the basic productivity strategies*: Zero breakdown losses, zero setup losses, zero tool replacement, etc.
- ■ *Level 2, the departmental productivity strategies*: Work measurement two times/year, continuous measurement of OEE for selected equipment, developing standards for new and existing equipment, developing training plans for each operator for increasing manufacturing process capacity, etc.
- ■ *Level 3, the product family productivity strategy*: Identify/locate and continually measure losses and waste for each process and set priorities for improvement based on CLW and CCLW for increasing manufacturing process capacity.

The strategy for annual MCI is developed depending on the sales for each PFC. If the sales are increasing, then the reduction of transformation costs will be monitored predominantly, in particular of manufacturing overheads, by increasing in particular the capacity of the equipment. If the sales are declining, then the reduction of direct and indirect labor costs and of direct and indirect material costs will be a priority. The basic strategy of annual MCI is based on the continuous determination of CLW and CCLW.

After these detailed analyses for each PFC for the identification of productivity and costs strategies based on the needs for reducing unit manufacturing costs and weaknesses, the annual MCI targets and means may be set for meeting the annual MCI goal.

4.1.2 Step 2: Costs Strategy into Action—MCPD Alignment for Setting the Annual MCI Targets and Means

The MCPD system is first applied in the upper part of an organization and deployed down through every PFC/product/process/activity and every function and section, or it can be applied "stand-alone" within a department.

The principles and mechanisms are the same; the only difference is that the alignment of the efforts and resources is done at different levels.

The MCPD system is a process used at all the levels of an organization to align corporate, PFC, product, work center, process, activities, and targets with the overall direction and plans of the organization. In step 1, the external context was aligned to the internal context of the company for the following years and the basic strategies of productivity and MCI were defined. In this step, the external priorities are aligned to the priorities of the organization for the following year.

The visibility of the efficiency and effectiveness of the MCPD system, implicitly of the process of general alignment of the company by the annual MCI targets and means, is in the constant decline of transformation costs and of material costs for each PFC. The results of the MCPD system are visible in fulfilling the annual target profit by meeting the annual MCI goal (see Formula 1.2).

4.1.2.1 Reconciling Top-Down and Bottom-Up for Setting the Annual MCI Goal and Annual MCI Targets

From the perspective of the MCPD system, the continuous easing between the two apparently antagonistic parts, the exterior of the company and its interior, as a basis for the continuous and harmonious transformation, which might support the long-term profit plan (multiannual targets profit), is done by the process of annual reconciliation, or whenever need be, between the annual manufacturing target profit and the annual MCI targets to meet the annual MCI goal (see Figure 3.4). This reconciliation is one of the basic principles of the MCPD system: "The continuous reconciliation of the annual MCI targets for each product family cost" (see Principle no. 4, Table 1.1).

The key of the MCPD system, at any level of the company, is the *alignment of the annual MCI targets to the annual MCI goal.* The alignment process aims at PFC, product, work center, process, activities, and targets that are in decline in the company, and it must be correlated with the SKPMP, with the PCBG and with the OMIs, with the company productivity mission (CPM), and with the company productivity vision (CPV)— multiannual target profit, market share, and sales volume.

In this context, the purpose of the MCPD system is that every individual in the company, in his/her area of work, should have a clear understanding of the bottom-up and top-down relationships, more precisely of the connections between the annual MCI targets and annual MCI means (cascading the annual MCI targets and means to meet the annual MCI goal) for each PFC (see Formulas 1.4, 1.5, and 1.7), of the way of fulfilling the annual MCI means (see Formulas 1.8 through 1.10), and of the resources allocated, and the evaluation of the performances, everything in order to meet the MCI targets. This full visibility is supported by the participation of all the people in the company in the specific processes of the MCPD system, both in setting the annual MCI targets (catchball process and reconciliation) and especially in their fulfillment by annual MCI means (PST, kaizen, and kaikaku).

In the MCPD system, the process of annual strategic alignment is achieved by the three levels of annual MCI reconciliation (see Figure 3.4) by which the *reconciling top-down and bottom-up for setting the annual MCI targets* is achieved.

The annual reconciliation is required to set the *annual MCI goal* for the whole company, for all the PFC, and then to set the *annual MCI targets for every PFC* (see Figure 4.6) based on the annual MCI means identified and accepted.

Annual reconciliation (top-down, *market driven* [target price] and *profit driven* [annual manufacturing target profit], and bottom-up, *MCI targets deployment* [annual CLW and CCLW targets] and *MCI means deployment* [annual losses and waste targets and annual critical losses

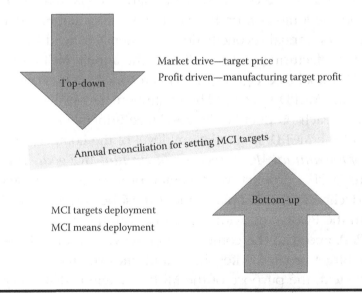

Figure 4.6 Top-down and bottom-up approach for the annual reconciliation of MCI targets and means.

and waste targets]) sets the annual MCI goal required to meet the annual profit target and, implicitly, the future state of MCI or the future level of CLW that may be approached by annual MCI means. The future required level of CLW (the required level of reduction of manufacturing costs affected by CLW) is set by senior managers to draw the future direction of the company from the perspective of annual MCI and implicitly of the annual MCI goal.

Further, *the annual MCI goal deployment* is the alignment of all the annual MCI targets at the level of the processes of each PFC (*bottom-up*) to the strategic need for manufacturing cost reduction (*top-down*). For example, if the annual MCI goal is of 6% per year, attempts will be made that each PFC might contribute both to meeting the 6%, by fulfilling the annual MCI targets in order to achieve the level of competitiveness by the products' price. Table 4.2 presents the future state of CLW for all processes of a PFC in order to contribute to the fulfillment of the MCI goal, together with other PFC—see continuation of Table 4.1.

So, the annual MCI goal deployment is fulfilled by starting from the need of annual reduction of manufacturing costs (top-down approach) and continues at the level of every process of a PFC by setting MCI targets and means. As can be seen in Table 4.2, the percentage of CLW improvement are set for every process of PFC, in CLW are also included CCLW, in order to meet the annual MCI targets, starting from the need for the annual MCI goal, for example, of 6% per year to meet the annual profit target.

When setting the annual MCI targets, special attention must be paid to the connections between cause and effect, in order not to tangle the cause with the effect (annual MCI targets are the expected effects; wanted managerial control on CLW and CCLW and the annual MCI means are the causes). Therefore, the annual MCI targets at the level of the processes of every PFC are set based on the annual MCI goal (top-down approach) and on the annual MCI means (bottom-up approach).

By this annual reconciliation, the pressure of the market (price and profit) is brought at the level of the objectives and realities corresponding to the manufacturing flow (product number increase, speed and agility increase, lead time decrease, productivity increase, and product quality ratio increase) with the help of the annual MCI means, to fulfill the annual MCI goal by the annual MCI targets.

Table 4.2 Cost of Losses and Waste: Current Stage vs. Ideal Stage vs. Future State for PFC Processes

	Current State (1)	CLW of Current State (2)	Ideal State (3)	CLW of Ideal State (4)	Future State (5)	CLW of Future State (6)
Production numbers	100%	$X	125%	$X-30%	105%	$X-6%
Waste: finished products stock	100%	$X	-98%	$X-92%	-8%	$X-30%
Process "n"						
Process 2						
Equipment losses (%, hours, and $)	44%	$X	8%	$X-35%	-7%	$X-15%
Human work losses (%, hours, and $)	25%	$X	1%	$X-35%	-5%	$X-20%
Material losses (% and $)	6%	$X	0%	$0	-2%	$X-35%
Energy losses (% and $)	4%	$X	0%	$0	-1%	$X-40%
Setup (minutes)	15	–	5	–	9	–
Total cycle time (TCT) (minutes)	3 min	–	1 min	–	2 min	–
Transfer time (seconds)	270	–	60	–	180	–
Number of workstations or operations	3	–	1	–	3	–
Takt time max/bottleneck (seconds)	22.5	–	5	–	18.5	–
Waste: WIP setup (WIP S) (minutes) or units	30 min	$X	5 min	$X-83%	15 min	$X-50%
Waste: WIP transfer (WIP T) (minutes) or units	5 min	$X	1 min	$X-80%	3 min	$X-40%
Process lead time (TCT + WIP S + WIP T) (minutes)	38 min	$X	7 min	$X-81%	20 min	$X-50%

(Continued)

Table 4.2 (*Continued*) Cost of Losses and Waste: Current-Stage vs. Ideal Stage vs. Future State for PFC Processes

	Current State (1)	CLW of Current State (2)	Ideal State (3)	CLW of Ideal State (4)	Future State (5)	CLW of Future State (6)
Process 1						
Equipment losses: %	45%	$X	8%	$X-35%	-7%	$X-15%
Human work losses: %	26%	$X	1%	$X-35%	-5%	$X-20%
Material losses: %	5%	$X	0%	$0	-2%	$X-35%
Energy losses: %	3%	$X	0%	$0	-1%	$X-40%
Setup (minutes)	20	–	5	–	15	–
TCT (*minutes*)	45 min	–	18 min	–	40 min	–
Transfer time (seconds)	600	–	30	–	500	–
Number of workstations or operations	31	–	10	–	28	–
Takt time max/bottleneck (seconds)	1,975	–	1,200	–	1,550	–
Waste: *WIP setup (WIP S) (minutes) or units*	20 min	$X	5 min	$X-60%	9 min	$X-50%
Waste: *WIP transfer (WIP T) (minutes) or units*	15 min	$X	5 min	$X-66%	9 min	$X-60%
Process lead time (TCT+WIP S+WIP T) (minutes)	80 min	$X	28 min		58 min	$X-50%
Waste: raw materials stock	100%	$X	-92.30%	$X-90%	-7%	$X-50%
Waste: components stock	100%	$X	-91.70%	$X-90%	-7%	$X-40%
Waste: packaging stock	100%	$X	-88.90%	$X-86%	-8%	$X-45%

The main characteristics of the top-down approach are as follows (Posteucă and Sakamoto, 2017, pp. 82–88):

- *One-way approach*: The annual MCI goal is stringent and coupled with the annual target profit and further with the long-term profit plan (multiannual targets profit, external target profit, and internal targets profit—from MCI); "Target profit from MCI does not change" (see Principle no. 1, Table 1.1).
- *Targets come from senior managers*: The annual MCI targets are for each PFC and are aligned to the annual MCI goal for the whole company; "MCI targets for each product family cost" (see Principle no. 2, Table 1.1).

The main characteristics of the bottom-up approach are as follows (Posteucă and Sakamoto, 2017, pp. 82–88):

- *Employees keep track of the annual MCI targets*: A specific, scientific, and analytical approach for each PFC; "The continuous quantifying of losses and waste in costs for each product family cost" (see Principle no. 3, Table 1.1); searching to uncover hidden reserves of profitability.
- *Several ways of approach for the annual MCI means*: The annual MCI goal is set by the senior managers and middle managers; implementation teams, and each person must meet the annual MCI means; "Coordination improvements through MCI targets for each product family cost" (see Principle no. 6, Table 1.1).
- *A strong and consistent impact on the manufacturing flow*: The tangible and intangible results of the MCPD system on the manufacturing flow are continuously monitored at the level of each PFC in order to permanently keep the annual direction for meeting the annual MCI goal; "Improvement budgets for each product family cost" (see Principle no. 5, Table 1.1).
- *A cause-and-effect approach for the annual MCI targets*: The reconciliation will aim at solving the root causes of every PFC; "Waste (stocks) elasticity on losses" (see Principle no. 7, Table 1.1).

So, the process of annual reconciliation or whenever need be, usually twice a year (in July of the current year and in January of the next year), between the annual manufacturing target profit and the annual MCI targets is done according to the modifications of the internal environment of the company (annual MCI means) and in particular to the modifications of its external environment (market driven, target price; profit driven, annual manufacturing target profit and annual external expected profit).

4.1.2.2 Setting the Annual MCI Targets and Means for Each PFC

After achieving the annual reconciliation and determining the necessary level of the annual MCI goal and of the annual MCI targets at the general level, or according to the managerial wishes for the annual MCI targets for each PFC, *the biggest challenge of the MCPD system is setting the annual MCI targets and means acceptable at the level of each PFC.*

Setting the annual MCI targets and means starts from the current state of CLW, from the ideal state of CLW, and in particular from the future state of CLW. Most of the time, the annual MCI targets and means are already set when the future state of CLW is designed to meet the annual MCI goal.

In order to have the annual MCI targets realistic and attainable within the time imposed by the market (the pressure on the price and profit), before setting the annual MCI targets and means, we must know already the following:

1. Actual losses and waste from processes and operations for each process of PFCs
2. The potential level of approaching CLW and CCLW based on the relations of causality between losses and waste along the manufacturing flow of each PFC and their impact on the whole manufacturing system
3. The losses and waste behaviors and associated costs—based on annual quarantine analysis of CLW and CCLW
4. Potential annual MCI means

Figure 4.7 presents an example of setting the annual MCI targets deployment for a PFC by its alignment to the need to meet the annual MCI goal, for example, of 6% per year in order to meet the annual target profit.

For example, the situation of Section "1A3c" will be commented. Thus

■ *In matrix 1*: The level of importance of losses and waste collected for every process of one PFC (PFC1) (physical losses and time-related losses) is presented (Posteucă and Sakamoto, 2017, pp. 119–122). In order to determine the level of importance, significant thresholds are set by means of Pareto analysis for the previous period (usually, the

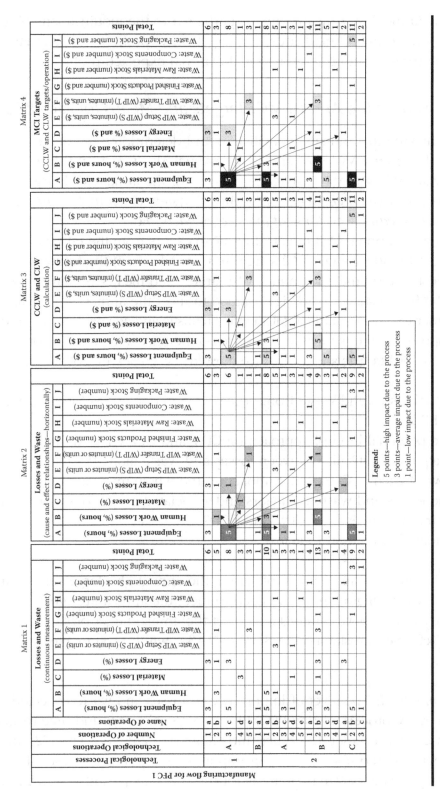

Figure 4.7   Matrices for the annual MCI targets deployment for a PFC.

last 12 months; the comparison is always made with a reference on the losses and waste level, which is updated for the following five years: benchmarking at group level, best practice, or ideal state; the reference considered at the time of annual drafting the master budget—to set the level of the annual external profit) and scenarios for the future evolution are developed, in particular for *SKPMP*. In the situation "1A3c," it is about the *equipment losses* (*breakdown losses*) noted with the level of importance 5 based on continuous measurements (breakdown rate of equipment, 1/hour; repair rate of equipment, 1/hour) and number of defects produced by the breakdown (unit) and energy *equipment losses* noted with the level of importance 3, based on the continuous measurements of consumption (theoretical and actual cost for "1A3c").

■ *In matrix 2*: The levels of importance of losses and waste based on the relations of horizontal causality are presented. Going back to the situation "1A3c," referring to the equipment losses (breakdown losses), the notation with the level of importance 5 is maintained, because it is considered that this is the cause of the effects on "1A2b," human work losses (motion/walking losses, man * hours/month), which decreases from the level of importance 3 to level 1, due to the fact that a part of the lost times of the operators are caused by the breakdown events; it is measured as man * hours/month. Similarly, it is approached with the decrease of the levels of importance for "1A3c" (energy equipment losses), "1A4d" (material losses; scrap ration, %), "1A5e" (waste: WIP transfer, minutes or units), "2A1a" (human work losses—man * hours/month), "2A3c" (equipment losses, %—hours), "2B2b" (energy equipment losses and waste—measured in minutes and/or units: WIP transfer—WIP T) (minutes or units), and "2C1a" (energy equipment losses); for technological processes "A," a part of losses are specific for the process and have no uncontrolled impact on the manufacturing flow. For example, "1A1a" that suffers because of the current standard that is too high for scheduling shutdown time is noted with level 3. This activity will remain noted with level 3 because an improvement of the current standard is required (lubrication standard time). Most of the losses and waste are collected by means of OEE. In the shift meetings (daily MCI management), every stoppage of the processes/equipment is analyzed and the relationships of causality of waiting times are set. This waiting time is separated between the waiting for the proper intervention on an equipment or process and waiting for another equipment or process in the flow. This information is written down in the change report. At the end of the month, the settlement of equipment and process non-productivity and their relationships of causality is precisely known.

■ *In matrix 3*: A level of importance from the perspective of CCLW and CLW is set based on calculation. The performance of CCLW and CLW is to identify a minimum 30%–35% of the manufacturing costs of PFC (both direct and indirect costs, as well as fixed and variable expenses). In the situation "1A3c," referring to the equipment losses (breakdown losses) previously noted with the level of importance 5, this level of importance is maintained because the breakdown event draws CLW from several activities of process 1 (further from process 2 and further from the manufacturing area) and is approached as being CCLW. CCLW associated to breakdown losses in process 1 draw a part of the costs corresponding to the activities that had the root cause level of breakdown losses and that were previously described in matrix 2. Instead, following the transformation of losses and waste in manufacturing costs, a part of losses and waste have a high level of manufacturing costs and need the setting of the annual MCI targets based on the subsequent annual MCI means. So, a part of losses and waste in process 1, first noted with level 1 in matrix 2, were noted with level 3 in matrix 3: "1A3c," energy losses; "1a5e," waste, WIP transfer. At the same time, certain losses and waste remain with the initial notation ("1A1a"—energy losses). This way, we can make probable scenarios for the annual MCI targets for each PFC (see Section 3.1.3.5).

■ *In matrix 4*: The annual MCI targets are set in order to meet the annual MCI goal (e.g., of 6% per year for PFC1), based on potential and feasible annual MCI means. As we can see, the activities that need setting the annual MCI targets were chosen, both for the activities generating CCLW ("1A3c"—*critical cost of breakdown losses*) and for those generating CLW, which are visible at the technological processes "1" and technological operation "A" level ("1A1a," cost of energy losses; "1A3c," cost of energy losses; "1A5e," cost of waste, WIP transfer). In order to meet the annual MCI targets for process 1, annual MCI means were identified as follows: (1) development of energetic management—MCI means level 3, *kaikaku* (a new method of management that approaches also manufacturing system for all PFC), (2) development of two kaizen projects to reduce energy consumption in "1A1a" and in "1A3c," and (3) initiation of a kaizen project for "1A5e," cost of waste: WIP transfer (MCI means level 2, kaizen). Moreover, the annual MCI targets and means were set for process 2. Both annual MCI targets and annual MCI means are advised to the whole company and explained precisely at the place where the improvement activities will be carried out.

Further, for every annual MCI target, annual MCI means (*action items to set MCI means*) are set. Figure 4.8 presents the annual MCI means levels 2 and 3 for the previous example. As we can see, seven *kaizen* projects were set for the following 12 months and three *kaikaku* projects (introduction of energetic management for the whole company and replacement of an equipment to cope with a new capacity imposed by the customers' demand and to decrease the unit manufacturing costs by increasing the productivity level). The following activities were considered as generating CCLW and had priority in planning the annual MCI means for approaching the following:

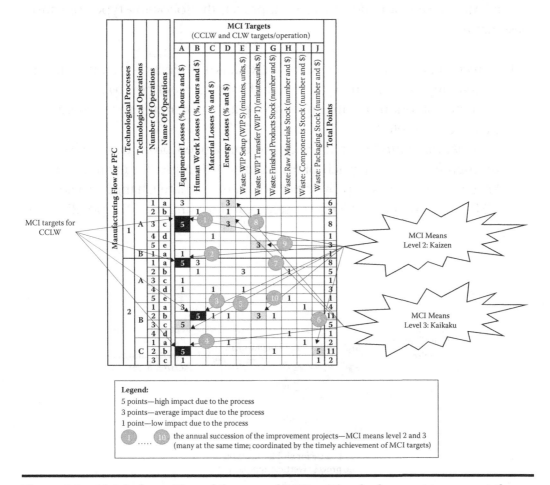

Figure 4.8 Matrix for setting of the annual MCI means deployment to support the annual MCI targets.

"1A3c" (equipment breakdown losses, kaizen), "2A1a" (equipment minor stoppages losses, kaizen), "2B2b" (human work losses/management losses; changes in production planning, kaizen), and "aC2b" (equipment cycle time losses; insufficient capacity of the equipment in terms of OEE of 84%, considered maximum and, even if it increased, could still not cope with the new capacity imposed by the market, and the effort for improving OEE is economically unjustified, kaikaku). The catchball process underlies the development of the annual MCI means deployment (targets for losses and waste). The annual MCI means deployment must be accepted by all the individuals involved in meeting the annual MCI targets for a PFC and, implicitly, to contribute to the annual MCI goal of the company.

In fact, the annual MCI means deployment aims at approaching all the types of causes described in Figure 4.4. So, from the perspective of the annual MCI means, it is necessary to approach the following types of causes (see Figure 4.9):

1. *Direct cause* (in particular through kaikaku projects—manufacturing system approach by new methods of management/control, for example, autonomous maintenance, planned maintenance, energy management, etc.)
2. *Contributing cause* (in particular through kaizen projects, but also through PST; the approach of an annual MCI target, for example, a kaizen project for setup by means of Single-Minute Change of Die (SMED) technique or MDC) (Posteucă and Sakamoto, 2017, pp. 257–377)

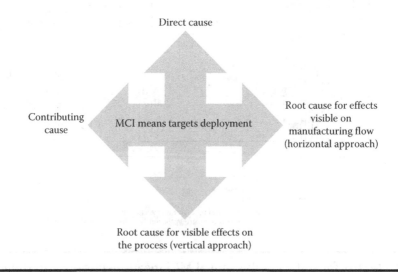

Figure 4.9 Annual MCI means targets deployment.

3. *Root cause for visible effects on the process* (in particular through kaizen projects, but also through PST; approaching the maintenance and improvement of an annual MCI target, e.g., a A3 project to eliminate the variations from the standard of the lubricants consumption for an equipment)

4. *Root cause for effects on the manufacturing flow* (in particular through kaizen and kaikaku projects; approaching an annual MCI target, e.g., a kaizen project by reducing or eliminating the breakdown for an equipment)

Starting from the annual scenario for MCI targets deployment, the annual MCI means deployment or *action items to set the annual MCI targets* for each PFC, for each department or section, at every stage of the manufacturing flow are developed.

Figure 4.10 presents the connection between the three levels of approaching the annual MCI targets deployment, each of the three levels being based on PDCA cycle:

■ *Means level 1: PST for manufacturing costs* (enhancing operational control on the annual MCI targets by solving associated problems, implicitly of annual CLW and CCLW targets to meet the annual MCI goal)

■ *Means level 2: kaizen projects to meet the annual MCI targets* (systematic improvement of the annual MCI targets level by kaizen projects, implicitly of annual CLW and CCLW targets to meet the annual MCI goal)

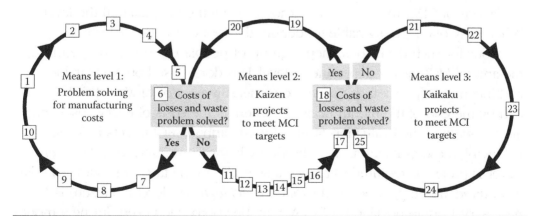

Figure 4.10   Annual MCI means deployment trilogy for each PFC.

■ *Means level 3: kaikaku projects to meet the annual MCI targets* (systemic improvement of the annual MCI targets by kaizen projects, implicitly of annual CLW and CCLW targets to meet the annual MCI goal)

Depending on the set priorities (see Section 3.1.4) for the three levels of approaching the annual MCI targets deployment, a relevant number of kaizen and kaikaku projects are set that might contribute to achieving the need to reduce the unit manufacturing cost at the level imposed by the market and to the achievement of the multiannual internal profit plan (by means of the annual MCI goal).

4.1.2.3 Annual MCI Means Deployment (Level 1): PST to Keep the Annual MCI Targets

The first level of the annual MCI means approaches the deviations from the annual MCI targets already set and focuses on the identification, elimination, and prevention of the variation of the current level of CLW and/or CCLW (acceptable current standard). So, the real reason that contributed to the creation of a problem concerning CLW or CCLW will be looked for, more precisely the localization in processes of costs with nonproductivity.

The annual MCI means deployment (level 1) is a continuous, proactive activity and not a deeply reactive one. The significance threshold of the annual MCI targets is set at a level that might prevent an advanced state of deviation of CLW and CCLW from the standard level for a certain period of time. Every manufacturing flow area has such a significance threshold defined.

Moreover, the procedure of rapid reaction in the case of significant deviations of CLW and/or CCLW is known by all the people at all the levels. When only one man is unable to identify the root cause and objective solutions for such deviations, then a group of people is formed. Triggering the annual MCI means deployment (level 1) is done based on measurable and incontestable proofs, by continuous measurement of CLW and CCLW and by the end-of-the-month decision, when all the costs of every cost center are known and the evaluation of meeting the annual MCI targets is made. Following these evaluations, it is decided whether a *MCI means deployment* (level 1) project is required to reduce a standard in the target or a new standard is required, and triggering the *MCI means deployment* (level 2, kaizen) is necessary. Certain deviations of CLW may be approached even during a month, without waiting for the quantification in costs of losses and waste.

So, the annual *MCI means deployment* (level 1, PST) refers to the enhancement of the operational control process on CLW (and on CCLW formed by summing up CLW along the manufacturing flow) and solving the associated problems. The main operational steps are as follows:

1. Measuring the current state of CLW and visual presentation of the measurement results to enable a rapid and easy recognition of CLW deviations from the standard level (visualization of the abnormal state).
2. Developing the clear and consistent communication system for every hierarchical level of the current stage of deviations regarding CLW and the organization of effective and efficient meetings (usually every morning between 8 and 15 minutes; setting a short duration of the meetings is to determine and document beforehand with data and real facts all the topics to be discussed).
3. Determining the priorities and focusing on the top CLW problems (usually top 5).
4. *Go and see* for accurate understanding of the current state of CLW (On-The-Spot CLW Analysis; why the problem arose and what makes it grow).
5. Reconstitution of the standard conditions of CLW by means of a structured approach of PST for manufacturing costs (defining the problem, analysis of the problem/RCA, and consistent implementation of the chosen solutions for the remedy of the root cause) that should support continuous learning, teamwork, and transparency—using in particular A3 technique.
6. Verifying whether the level of CLW was reduced to the standard level or whether the principles and phenomena at the process level, which caused the CLW, were reduced or eliminated; SOP and workplace standards are verified.
7. *If so*, then the operational control process on CLW is optimized not to avoid overtaking CLW; *if no*, step (5) will be resumed or a *kaizen* project will be initiated for SOP restoration.
8. Setting the monitoring interval of CLW level for a certain period of time, self-controlling and reminder.
9. Setting a reaction plan in the case of significant deviations from the targets corresponding to CLW.
10. Extending the solutions, methods, techniques, and social skills of the people resulted from the problems concerning CLW (by means of one-point lesson, OPL).

4.1.2.4 Annual MCI Means Deployment (Level 2): Establishment of Kaizen Projects to Meet the Annual MCI Targets

The annual MCI means deployment level 2 refers to the improvement of the current level of CLW by kaizen projects to set new standards for CLW within a maximum of six months. As mentioned earlier, a kaizen project focuses on a certain type of CLW and is initiated when *MCI means deployment* (level 1) was unable to bring a standard back to the target values, when the improvement of a standard is imperative or when a standard is developed for the first time (SOP). *If after following step (6) mentioned earlier (level 1) the answer is negative*, that is, the CLW level was not reduced to the standard level, using *PST for manufacturing costs*, then the following steps of the *kaizen* project for CLW will be made:

11. Identification of the types of CLW that require definition and/or redefinition of the standard.
12. Formation of interdepartmental team that will approach CLW.
13. Detailed definition of phenomena, principles, and parameters that contribute to the nonobservance of the current standard of CLW and that impose a new standard—defining the problem.
14. Setting an acceptable target of improvement of the current state of CLW (measured in step 1 in level 1).
15. Detailed development of an action plan for a maximum of six months to achieve a new standard of CLW.
16. Detailed analysis of the problem regarding CLW (RCA, vertically) and setting the root cause.
17. Defining the required actions and measures to remedy the root cause (temporary and permanent) and determining the necessary resources.
18. Verification of the results obtained regarding the fulfillment of the improvement targets of the current state of CLW, after implementation of actions and measures; *if CLW target was not met*, then work will be resumed from step 16 (detailed analysis of the problem regarding CLW) or a *kaikaku* project will be initiated.
19. *If the CLW target was met*, then the new standards for CLW are defined to prevent the recurrence of the same problem and define the benefits following the cost/profit analysis.

20. Defining the plans of horizontal extension of the new standards for CLW for similar problems that may be approached by means of the solutions identified and plans for the future resulted from CLW improvement, identified during the kaizen project for CLW (by means of OPL).

After step 20, it will be continued with steps 7, 8, 9, and 10 of *MCI means deployment* (level 1). *The earlier mentioned steps are made both to set the standards for CLW and also for the kaizen strategic projects to set the standards for CCLW.* Examples of such kaizen project can be as follows (see Formula 3.9):

a. *For variation of standard cycle time of bottleneck*: Stabilizing and reducing cycle time; line organization/balancing, etc.

b. *For variation of process capacity*: Reducing breakdown time, reducing equipment minor stoppages time, reducing scrap and rework, improving energy consumption, improving the consumption of auxiliary materials and raw materials, PM—analysis for defects and breakdown, etc.

c. *For variation of standard setup time*: Stabilizing and reducing setup time; 5S for setup; MDC for setup, improving locations on the layout, improving measurement and adjustment times, etc.

d. *For variation of standard transfer time*: Reducing non-value-added motion/walking time, reducing material handling time, 5S in the dedicated zones for WIP, etc.

4.1.2.5 Annual MCI Means Deployment (Level 3): Establishment of Kaikaku Projects to Meet the Annual MCI Targets

They refer to the improvement of CLW and CCLW level by planning *kaikaku* projects and implementing corresponding solutions, usually for a period of time of more than one year, because they are often considered strategic multiannual projects. They are initiated when the results of kaizen project in step 18 as previously mentioned are not sufficient (*MCI means deployment*, level 2) and are unfeasible, the use of new managerial methods is imperative for strategic reasons, or new equipment/technologies/processes are required. The main steps of a *kaikaku* project for CLW or CCLW are as follows:

21. *Concept planning*: Defining the new concept for innovation and/or the new managerial method that should approach CLW and/or CCLW improvement.
22. *Action plan*: Defining the necessary action plan.
23. *Implementation of the new concept*: In the first stage within a pilot project (if possible).
24. *Testing the results of the pilot project*: Verification in detail of the different potentials between the targets set for CLW and/or CCLW and the results obtained from a pilot project.
25. *Extension plan*: According to the action plan, the horizontal and vertical extensions will be done to realize the benefits of implementing *kaikaku* projects for CLW and/or CCLW (by means of OPL).

After step 25, it will be continued with steps 19 and 20 of *MCI means deployment* (level 2) and then steps 7, 8, 9, and 10 of *means level 1*.

Examples of *kaikaku* projects for CLW and CCLW can be as follows:

a. *For innovation*: Products, equipment, processes, and technologies
b. *For new managerial methods*: Annual manufacturing improvement budget (AMIB) for existing products, multiannual manufacturing improvement budget for new products, annual manufacturing cash improvement budget (AMCIB), daily manufacturing cost management, departmental organization for achieving the annual MCI targets, autonomous maintenance, planned maintenance, etc.

The MCI means trilogy deployment for each PFC will underlie the manufacturing cost policy development (second phase of the MCPD system), namely the development of annual improvement budgets (step 3 of MCPD) and the setting of annual projects to meet MCI targets (step 4 of MCPD). Further, also depending on MCI means deployment trilogy for each PFC, the *manufacturing cost policy management* (third phase of the MCPD system) will be developed and *the structure of necessary people to approach the* annual MCI means (step 5 of MCPD), the method of evaluation of the performances of the annual MCI means (step 6 of MCPD), and the organization requirements at the shop floor level for daily MCI management (step 7 of MCPD) will be determined.

4.1.2.6 *Communicating the Annual MCI Targets and Means*

The purpose of the annual communication of MCI targets and means is to improve the effectiveness of communication and reaction by the rapid transmission of information at every hierarchical level through the catchball process.

The internal communication of the annual MCI targets and means is part of visual management at the shop floor level management. By this communication, the current and future states of CLW are presented for every process of each PFC. At the same time, the mode of recognition, investigation, and elimination of the deviations from the current state of the annual MCI targets is determined. Then, the annual communication of MCI targets and means becomes a continuous communication based on the priorities in approaching the problems regarding the annual MCI targets; the continuous measurement of losses and waste and continuous calculation of CLW and of CCLW are continuously determined.

An effective communication of the annual MCI targets and means contains the following:

- CPV
- CPM
- PCBG
- Annual SKPMP
- Annual MCI goal for each PFC
- Annual MCI goal for all PFC (of plant)
- Annual MCI targets—annual CLW/CCLW targets for each process of each PFC
- Annual MCI means (inclusive of the annual MCI means targets) for every annual MCI target
- Annual MCI action plans (start and finish dates for kaizen and kaikaku projects)
- Assigned resources for each action (kaikaku) and activity (kaizen)
- Responsible managers for each improvement project together with the types of KPIs to be improved by the annual MCI means
- Annual organization of departments for the annual MCI means— the lists of persons who will participate in each project of improvement
- Annual training plans

- Method of evaluation of the annual MCI means (in particular the connection with annual improvement budgets, with unit manufacturing costs, and annual master budget)
- Meeting structures for MCI means
- The persons inside the company who have access to these information
- Company logo and duration of the issue for the company

The information communicated to every person involved in the annual process of transformation of the company by MCPD must be simple and clear and at the level of comprehension of those who are supposed to meet the annual MCI means targets. An example of such a message sent on December 31, 2016, for a type of MCI means is as follows:

Do what: Improve

To what: CLW from the manufacturing flow for PFC "1"; technological processes "1" (assembly), technological operations "A," number of operations "4," name of operations "d" (setup)

How much: From 30 to 15 minutes or less (50%) (MCI means target); annual CLW are $135,000—optional (the stake in money of the MCI target)

By when: August 11, 2017

In conclusion, the total involvement of the top management is vital in order to engage the workforce to achieve the annual MCI targets and means.

4.2 Manufacturing Cost Policy Development

The purpose of the manufacturing cost policy development, the second phase of MCPD, is to coordinate the MCI means to meet MCI targets, with the aim of supporting MCPD in the long run and especially annually, by developing *AMIB for existing products, multiannual manufacturing improvement budget for new products, and AMCIB* (*Step 3: Annual Manufacturing Improvement Budgets Development*) and by planning and implementing the systematic improvement activities (MCI means level 2, *kaizen*) and actions of systemic improvement (MCI means level 3, *kaikaku*), both at the level of every PFC and at the interdepartmental level (*Step 4: Annual Action Plan for MCI for Each PFC)* to achieve *total and continuous involvement of all departments and beyond and reduction or elimination of reactive managerial behavior.*

4.2.1 Step 3: Annual Manufacturing Improvement Budgets Development

The coordination of external and internal opportunities of manufacturing companies, by continuously developing *productivity vision and mission* statements, PCBG, SKPMP, and strategies of productivity, has as central point the fulfillment of a multiannual target profit. In the MCPD system concept, the multiannual target profit is obtained by fulfilling the multiannual external profit expected from sales and multiannual internal profit expected from MCI.

Further, in one year, the annual target profit is accomplished by meeting the annual external target profit expected from sales and by meeting the annual MCI goal, both at the level of a group of companies and at the level of a plant in the group and at the level of a PFC of a plant.

In this context, in order to accomplish the annual target profit, we go beyond the annual external profit expected from sales. Annual external profit expected from sales is the traditional way of planning the annual profit through the budget network. In general, budgeting is the accounting instrument used to plan and control the activities of a company in order to ensure the fulfillment of the annual strategic plans focused on sales. These budgets have the purpose of monitoring the financial results of a plan for a certain period of time, usually 12 months, sometimes with their continuous restoration (rolling budget). They represent the quantification of the operational and financial decisions, both in the planning phase and in the result control phase.

In the logic of the MCPD system, in order to fulfill the annual external profit, it is necessary to develop the annual master budget, both of *operating budget* (selling budget/revenues budget, cost of sold goods budget with direct material purchases budget, with direct labor budget, and with manufacturing overhead budget and of selling and administrative expenses budget), of *financial budgets* (capital expenditure budget, cash budget), and of *pro forma financial statements* (budgeted income statement and budgeted balance sheet). From the perspective of the MCPD system, which focuses on the *operating budget* (except for the selling budget), with an extension over *cash budget*, every structure of the *operating budget is a CLW and CCLW bearer.*

At the same time, in the logic of the MCPD system, in order to meet the *annual MCI goal*, it is required to identify CLW and CCLW in the structure of operating manufacturing budget for every process of every PFC and/or product and per total plant. The *annual improvement budgets* are developed for this with the aim of transposing the strategic planning of internal profit in the annual MCI goal and further in the annual MCI targets and means and in annual MCI action plans. This way, the employees are motivated to fulfill the annual MCI goal having instruments for visualization of the differences between the annual MCI targets and the current state of the manufacturing expenses.

Therefore, the manufacturing companies transformation through MCPD system to meet the multiannual targets profit will aim both at external profit expected from sales and at internal profit expected from MCI (by uncovering hidden reserves of profitability).

4.2.1.1 Annual Manufacturing Improvement Budget Framework

It is aimed by means of *annual manufacturing improvement budgets* (*AMIB*) to observe Principle No. 1 of the MCPD system ("Target Profit from MCI Does Not Change"). The observance of this principle (fulfillment of the annual MCI goal) together with the observance of the annual external profit expected from sales leads to the fulfillment of the annual target profit.

Even during the year external and internal events occur, which affect the planning of the annual target profit, the annual target profit level must be met at the end of the year. The two annual corrections of the MCI goal and of MCI targets and means are made (in July of the current year and in January of the next year).

Just like a company organizes itself minutely to obtain the annual external profit, by creating value to the products and services sold outside the company, it should organize similarly to obtain the annual internal profit (or annual MCI goal), in order to create a value product "sold" inside the company, which is called the MCPD system.

If MCI is the approach of unit manufacturing costs at the product and/or PFC level, *annual improvement budgets* are the approach of all PFC of the company with the aim of planning and controlling the level of the annual MCI goal.

Although the process of setting the annual MCI targets is a delicate one, because it implies the continuous coupling to the changing needs of the market (especially of the price) and at the level of the annual MCI goal, the major challenge of MCPD is at the level of determining the annual MCI means because it involves obtaining measurable results at the level of the processes of each PFC and at the required time, in fact, the future transformation plan of the manufacturing company is decided. In order to ensure the annual MCI targets and means success at the level of every process of PFC, the development of *control items* and implicitly of *control methods* is required.

From the perspective of MCPD, the development of the concept of *control items*, described by Kaoru Ishikawa (1962) (Ishikawa, 1980; Akao, 1991), and of the concept of *management by objectives*, also known as management by results, described by Peter Drucker (1954) (Drucker, 2010), implies the development of a method of control for every PFC and for the whole company for the following:

■ *Verification of the current results* (of the annual MCI targets and implicitly of actual manufacturing cost): *Control items* and *control points* are set for every structure of manufacturing costs (*control characteristics*); as it can be seen in the example in Table 1.3, there are
  – *Control items* is the whole detailing of *manufacturing costs on each PFC or product* (OMIs) for the seven levels of detailing of KPIs (transformation costs, direct labor costs and manufacturing overhead costs, equipment manufacturing overhead costs, etc.) and detailing of *total CLW for transformation costs and for material costs* (between 30% and 35% of total annual manufacturing costs on each PFC or product).
  – *Control points* are represented by *annual weaknesses of total CLW for transformation costs and for cost of material costs*, per levels of importance (approximately 20% of the total annual manufacturing costs on each PFC or product—those considered relatively easy, approachable, and feasible to meet the annual MCI goal) and of % *of target touch* (% *TT*) *of the annual MCI targets* (for *annual CLW targets* and *annual CCLW targets*) (approximately 5%–6% of total annual manufacturing costs on each PFC or product, those that are required to be approached to meet the annual MCI goal).
■ *Verification of current causes* (of the annual MCI means and implicitly of the causative factors, annual CCLW and annual CLW, and of annual losses and waste): *Annual inspection items* and *annual checkpoints*

or *annual inspection points* are set for every project of improvement (annual MCI means targets), for every annual CCLW means targets and annual CLW means targets, and in particular for every losses and waste. Going back to the example in Table 1.3, we have the following:

– *Inspection items* represent the *annual MCI means impact* per levels of importance (of the annual MCI means level 1, PST; level 2, *kaizen*; and level 3, *kaikaku*), on annual CCLW and on annual CLW and implicitly on annual losses and waste associated (e.g., reduces production stoppage due to the lack of material) (for the 5%–6% of total annual manufacturing costs on each PFC or product, chosen to meet the annual MCI targets by the annual MCI means targets).

– *Inspection points* represent % *of the annual MCI means target touch* (% *MTT*), implicitly of the process/equipment that is affected by losses and waste (e.g., production stoppage due to the lack of material) (for the 5%–6% annual MCI targets, by annual CLW targets and annual CCLW targets, chosen to meet the annual MCI targets by the annual MCI means targets, by annual losses and waste targets).

So, the control method developed in this respect is *annual improvement budgets* for the existing products, for the future products, and for cash flow management corresponding to the improvements in the manufacturing areas (Posteucă and Sakamoto, 2017, pp. 157–202).

Figure 4.11 presents *manufacturing improvement budgets cycle framework* in four big phases:

■ *Phase 1: Planning items to set the annual MCI targets.* All the activities from the productivity vision and mission statements to the annual MCI targets and annual CLW and CCLW targets

■ *Phase 2: Action items to set the annual MCI means.* The activities of setting the annual MCI means and annual losses and waste targets

■ *Phase 3: AMIB.* The activities of setting the (1) AMIB for existing products, (2) multiannual manufacturing improvement budget for new products, and (3) AMCIB

■ *Phase 4: Action items for annual action plan (execution).* Annual actions (kaikaku) and activities (kaizen) for MCI (identifying and implementing solutions to meet MCI targets for each PFC)

■ *Phase 5: Monitoring and evaluation of manufacturing improvement budgets.* Monitoring and evaluating the efficiency and effectiveness of

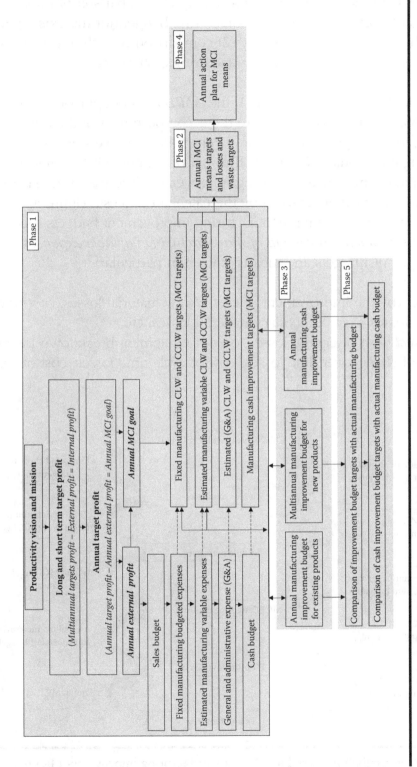

Figure 4.11 Manufacturing improvement budgets cycle.

manufacturing improvement budgets by (1) comparison of improvement budget targets with actual manufacturing budget (for the existing and for the future products) and by (2) comparison of cash improvement budget targets with actual manufacturing cash budget

In the logic described in Figure 4.11, the *AMIB* are not approached as a result, but as a purpose, to ensure the accomplishment of the annual MCI goal, for example, 6% per year, for the following five years, obtained by meeting the annual MCI targets through the annual MCI means. The thinking paradigm of manufacturing improvement budgets cycle has two ways of communication enabled by the catchball process (Figure 4.12).

The way of approaching manufacturing improvement budgets focuses on *the continuous improvement of manufacturing variable expenses* (by fulfilling MCI targets and means, kaizen in particular):

■ *Direct* (by the improvement of cost of raw material losses, cost of direct labor losses, cost of auxiliary material losses, etc.)
■ *Indirect* (cost of energy losses, cost of equipment breakdown losses, cost of changeover time losses, cost of cycle time losses, cost of scrap

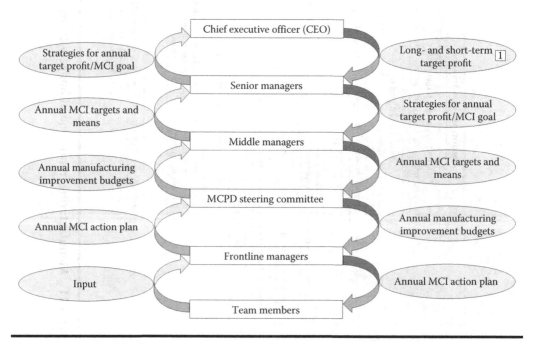

Figure 4.12 Catchball process for setting manufacturing improvement budgets.

losses, cost of rework losses, cost of maintenance material losses, cost of fuel losses, cost of raw material stock/stock days, waste, cost of finished products stock/stock days, waste, etc.)

The percentage of variable expenses differs depending on the industry and depends especially on the level of depreciation costs and indirect labor costs—*manufacturing fixed expenses.*

A matrix type organization and numerous cost centers are required in order to achieve a consistent cost management, with many KPIs for CLW and CCLW. All the decision on manufacturing improvement budgets are made at the interdepartmental level to increase the degree of decision-making objectivity.

The top managers develop manufacturing improvement budgets for the existing products and for the future products one year in advance (rolling budget), with the other part of the operational budgets for the achievement of the production. This way, time is allocated for scenario simulations regarding the annual MCI targets and means for the achievement of the annual MCI goal and supporting annual target profit.

Starting from the vision and mission statements of the company for 5–10 years, from the sales volumes, market share, and the level of target profit, a sales council sets the sales objectives for the following three to five years. Manufacturing improvement budgets aim at meeting the annual targets profit by successive simulations of the annual external profit from sales (volumes and prices) and of the annual MCI goal (annual MCI targets and means). Setting the annual MCI targets will aim both at ensuring the sales volumes (of the required capacities) and at improving the level of manufacturing expenses (in particular variable expenses—with great care on indirect variable expenses). The difference between annual targets profit and annual external profit estimated based on the sales volumes (usually smaller than the target profit) must be ensured by the annual MCI goal by fulfilling in time the annual MCI targets and means, because the increase of external profit by increase of prices is often hard to get, and the increase of sales volumes and, implicitly, of the production generates production costs that sometimes may be difficult to control on the background of a potential overstretching of current capabilities (overstretching that usually causes important percentage of losses and waste—with their associated costs). This is the reason why a part of the difference between the annual targets profit (e.g., 10% per year of the turnover) and the annual external profit estimated (e.g., 6.5% per year) is approached both by those who deal with sales increase and by the MCPD steering committee. If the sales volumes have an acceptable predictable upward trend, then this difference is divided into equal parts (50% at sales and 50% at MCPD). If the sales tend to drop significantly, then the cost reduction (or annual MCI goal) will take over more from this difference (e.g., 65%).

In order to improve variable expenses for meeting the annual MCI goal, it is necessary to identify CLW and CCLW corresponding to actual manufacturing expenses of the processes of every PFC to an approximate percentage of 30%–35%, a percentage where the approachable annual percentage will be taken of approximately 20% and the percentage that needs reduction, usually about 6%, choosing the most efficient and effective annual MCI means at the level of every process of each PFC. The estimation of variable expenses associated with component expenses originating from supplies (their price and quantities) is decided outside manufacturing and does not entail losses (except for costs of material handling losses—if applicable). They entail only waste: cost of components stocks/stocks days and costs of WIP. For them, the MCI targets and means will aim only the reduction and/or elimination of components stock days and useless WIP. So, the concentration zone of MCI targets and means is especially for the remainder estimated variable expenses (of transformation and material costs—at their current standard values).

Setting the annual MCI targets and means is inseparable from AMIB. By AMIB, it is presented from time to time the current stage of the annual MCI targets and means and, implicitly, of the annual MCI goal, and the potential corrections of the annual MCI targets and means are decided. AMIB may be developed for every PFC and for every product and every process of PFC with the purpose of monitoring the difference between the annual target profit and annual MCI goal. The traditional approach of the budget is restricted to the financial aspects and cannot support MCI to reduce the manufacturing cost. AMIB is a part of budgetary planning and control system, but it is approached independent of it, especially independent of standard costs (budgeted manufacturing expenses are summing up the standard costs of products). Standard costs, both of the existing products and of the future products, are the input data for the determination of their level of contamination with CLW and CCLW. Due to the annual MCI targets and means, the level of standard costs is continuously improved. In fact, for each PFC, the annual MCI targets are set depending on the volume of products planned annually and on annual CLW targets and on annual CCLW targets corresponding to manufacturing standard costs for each product and summed up for all the products of that PFC.

By AMIB it is set from the outset both the annual MCI goal level and the target annual cash generated by the fulfillment of the annual MCI means or the difference between the resources consumed for improvements (transformed into money) and the gains generated by meeting the annual MCI means targets (transformed into money—reduction of direct and indirect labor cost, reduction of manufacturing overhead costs, reduction of material costs, reduction of stocks or measured in reduced times, increased production capacity and avoiding the investment in new capacities, reducing the distances covered by people and/or product, reducing the length of the assembly line, reducing people's moves, and improving the morale of the people by increasing their extent of participation in meeting the annual MCI means targets). So, the role of AMCIB is to support the annual stake of MCPD, respectively of the annual MCI goal, by evaluating in money and gains obtained as a result of meeting the annual MCI means targets.

The focus on the annual MCI targets and means based on annual CLW and on annual CCLW, determined with high accuracy, makes the *MCPD a system of scientific steering of kaizen* (every day operating procedures—SOP) *and kaikaku projects* (new technologies, new equipment, new methods of management, etc.).

4.2.1.2 Steps for Development of the Annual Manufacturing Improvement Budget for Existing and New Products

The activities of setting the AMIB for existing and new products are carried out at the same time with setting MCI targets and means for every product and/or for every PFC and with annual action plan for MCI for every PFC, in view of fulfilling the annual MCI goal and, implicitly, the annual targets profit. The existing products are those products that were already launched in production and are in various stages of their life cycle (introduction, growth, maturity, and stabilization or decline). The new products are those products that are in various stages of development and will be launched in the next period and for which it is necessary to identify the sources for MCI with a view to recovering a potential part of the target profit lost throughout the life cycle of the product, especially in the first months after launching (the difference of target profit from the moment of acceptance of the new product and the moment of launching the new product, especially for the products with a long time to market, in the context of the decrease of the level of acceptable price and/or the increase of the raw material or of the components for suppliers' price).

Further, we will present an example of achievement of an *AMIB for the existing and for the new products of* PFC1 at a manufacturing company ("Plant A") in the manufacturing and assembly industry, for the year "N + 1."

In order to achieve the *AMIB*, it is necessary to have already the data and information of the annual entry:

1. Vision and mission statements of long-term productivity. At "Plant A" it was planned to increase by 15% the annual production volume for the following 10 years and to be among the first three global players; the multiannual targets profit for PFC1, in order to support an acceptable level, is $10 million for the following three years (annual operating targets profit obtained from the annual external profit and from the annual internal profit/annual MCI goal).

2. Defining PCBG: Manufacturing costs decrease, product number increase; speed and agility increase, quality increase.

3. Defining the basic strategies of productivity (strategy to increase the annual volumes of production based on the increase of the use of current capacities by investment).

4. Evaluation of past results and performances (analysis of the last three years evolution) for OMIs of KPIs and of kaizen and kaikaku indicators (KKIs)—*market driven for MCI, profit driven for MCI, driven by MCI targets deployment,* and *driven by MCI means deployment.*

5. Analysis of the internal situation of the last three years (SKPMP).

6. Analysis of external data for the last three years: Sales trend for each PFC and component product, price trend for each PFC and component product (price benchmark analysis; see Section 2.4.1), evolution of the competitors' capacities for each PFC, evolution of the launching rate of new products of the competitors, evolution of prices and of quality practiced by the suppliers (raw materials, utilities, etc.), evolution of inflation for certain zones of interest in the world, etc.

7. Planning the volumes of production for the year "N + 1."

8. Completion of the *budget committee's* structure (1, for annual external profit expected from sales, and 2, annual MCI goal expected from MCI, coordinated by the *MCPD steering committee; see Section 3.2.2.2*).

9. Determining the processes corresponding to PFC1 (see Section 3.1.3.2).

10. Determining the specific consumptions of every process of PFC1 depending on the mix of planned products for the following years and the production volumes (completion of the standard cost structures per cost centers for the year "N + 1").
11. Distribution of fixed expenses for every process of each PFC1, per cost centers for year "N + 1."
12. Distribution of estimated variable expenses for every process of every PFC1, per cost centers for the year "N + 1."
13. Identification of estimated variable expenses corresponding to components for every process of every PFC1, per cost centers for the year "N + 1."
14. History of the last 12 months of the measurements regarding losses and waste for every process of PFC1 and per total company, based on KPIs for losses and waste (Posteucă and Sakamoto, 2017, pp. 119–123) (see Figure 4.7).
15. Identification of the manufacturing key point forming critical losses and waste for PFC1 (see Section 4.1.1.1).

So, the steps of achievement of *AMIB* for PFC1 at "Plant A" for the year "N + 1" are as follows:

1. *Defining* the *total annual operating targets profit* (*see Table 4.3; row J*): $7,570,000 + $2,868,750 + $4,300 = $10,443,050.
2. *Defining annual external profit from sales* (*see Table 4.3, row H and rows A, B, and E*): $7,570,000 (a part of the total annual operating targets profit).
3. *Defining the annual MCI goal* (*see Table 4.3; rows C3 and F*): The annual MCI goal is determined in order to meet the annual target profit of minimum $10 million, starting from the level of the *annual external profit from sales of* $7,570,000. The annual MCI goal is $2,873,050 ($2,868,750 from the annual MCI target based on annual MCI mean targets accepted in the company based on annual CLW targets and annual CCLW targets at the level of the *manufacturing process* of PFC1, plus $4,300 of the total annual CLW targets and total annual CCLW targets identified and approachable for PFC1 process from *administrative expenses*).
   Note: For the determination of the annual MCI goal, the following big steps were made.

Table 4.3 AMIB Synthesis for Existing and Future Products for a PFC

Annual improvement budget calculations	Sales	Sales price ($) (a)	Annual External Profit Expected	Annual MCI Goal	Basic cost policy: Maximum annual operation targets profit
		$100	0	
	Sales quantities (b)	$850,000	0	
	(A) Sales revenue (a) * (b) ($)	$85,000,000	0	
	(B) Cost of goods sold budget (COGS) ($)	$77,000,000	0	
	(B1) Uncontrollable costs: fixed manufacturing budgeted expenses	$500,000	0	
	(B2) Controllable costs: estimated manufacturing variable expenses	$76,500,000	0	
	(B3) (From which) Uncontrollable costs: estimated manufacturing components expenses	$19,125,000	0	
	(C1) Controllable costs: total annual CLW and CCLW (identified in PFC1 process)	0	$17,212,500	
	(C2) Controllable costs: total annual CLW and CCLW (identified and approachable in PFC1 process)	0	$11,475,000	

(Continued)

Table 4.3 (*Continued*) AMIB Synthesis for Existing and Future Products for a PFC

	Annual External Profit Expected	Annual MCI Goal
(C3) Controllable costs: annual MCI target (based on annual MCI mean targets accepted in company)	0	$2,868,750
(D) Operating income (B) − (A) ($)	$8,000,000	0
(E) Sales and administrative expenses ($)	$430,000	0
(F) Controllable costs: total annual CLW and CCLW identified and approachable for PFC1 process	0	$4,300
(G) Total operating expenses (B) + (E) ($)	$77,430,000	0
(H) Operating income/annual external profit from sales (A) − (G) ($)	$7,570,000	0
(I) Operating margin (H)/(A) (%)	8.91%	0
(J) Annual operating targets profit (H) + (C3) + (F) ($)	$7,570,000 + $2,868,750 + $4,300 = $10,443,050	0
(K) Business tax (J) * 16% ($)	$1,670,888	0
(L) Net profit (J) − (K) ($)	$8,772,162	0

4. *Developing the control items per process of each PFC1 (verification of current state)*:

■ *Defining manufacturing uncontrollable expenses (Fixed Manufacturing Budgeted Expenses) (see Table 4.3, row B1)*: $500,000; in general, these expenses cannot be controlled directly by standardization but can be influenced at the level of unit manufacturing costs by improving and by reducing and/or eliminating costs of losses and waste behind them (such as depreciation expenses, indirect labor expense). For example, a cost with the depreciation of equipment cannot be reduced in one month. It is a fixed one. But, for example, the use of the equipment may be increased by increasing its availability by reducing the breakdown time, in order to reach a lower unit manufacturing cost. Moreover, the following uncontrollable manufacturing expenses were considered: depreciation costs, external logistic costs, financial costs, administrative costs, cost of rents in the manufacturing area, insurance costs in the manufacturing area, and other general and administrative expenses (*basic cost policy for uncontrollable expenses: absolute minimum costs*).

■ *Defining manufacturing controllable expenses (estimated manufacturing variable expenses) (see Table 4.3, row B2)*: $76,500,000; these expenses are controllable by standardization and by improvement, less *estimated manufacturing components expenses*: $19,125,000 (*see Table 4.3, row B3*; they have the price imposed by the suppliers). They find themselves at the level of every process of PFC1 and consist of the following:
  – Transformation cost (variable costs): DLC, direct labor costs; EMSC, external maintenance services costs; RMC, repairs and maintenance costs; DJC, die and jig costs; TC, tool costs; UC, utilities costs (*basic cost policy: minimum costs*)
  – Material costs (variable costs): DMC, direct material costs; IMC, indirect material cost (*basic cost policy: absolute minimum costs*)

■ *Determining the total annual CLW and CCLW (identified in PFC1, on cost center/process level), or a total annual CLW and CCLW for transformation costs and material costs (see Table 4.3, row C1)*: $17,212,500 (approx. 30% of B2–B3) (see Figure 4.7)
  – Transformation costs vs. CLW:
    • DLC vs. CLW (of equipment downtime losses; equipment performance losses, equipment quality losses; human work losses; yield loss)

- Indirect labor costs vs. CLW (of equipment quality losses, replanning loss; human work losses, management loss; yield loss)
- Manufacturing overhead vs. CLW:
  - RMC, repairs and maintenance costs, vs. CLW (of equipment downtime losses, breakdown loss; equipment performance losses, minor stoppages; inventory loss; yield loss)
  - UC vs. CLW (of equipment performance losses, minor stoppages, cycle time (speed down); equipment quality losses, scrap loss and rework loss; human work losses, measurement and adjustment loss; energy loss; yield loss)
  - TC vs. CLW (of human work losses, motion loss; energy loss; consumable loss)
  - DJC vs. CLW (of yield loss)
- Material costs vs. CLW:
  - DMC vs. CLW (of equipment downtime losses, setup and adjustment loss; equipment quality losses, scrap loss, rework loss; human work losses, adjustment loss; yield loss; raw material stock, waste; components stocks, waste; raw material and components WIP stock, waste)
  - IMC vs. CLW (of equipment quality losses, scrap loss, rework loss; consumable loss, yield loss; indirect material stock, waste; indirect material WIP stock, waste; packaging stock, waste)

■ *Determining the total annual CLW and CCLW identified and approachable for PFC1 process for administrative expenses (see Table 4.3, row F)*: $4,300 (approx. 1% of E)

5. *Developing control points in every process of PFC1 (annual weaknesses for CLW and CCLW: total annual CLW and CCLW, identified and approachable in PFC1 process) (see Table 4.3, row C2)*: $11,475,000 (approx. 20% of B2–B3) (see Figure 4.7):

   a. *Identifying annual weaknesses* (per levels of importance): For the processes that have significant and illogical variations of allocation of *manufacturing controllable and uncontrollable expenses, identified and approachable in PFC1 process*

   b. *Identifying annual weaknesses* (per levels of importance): For the processes that have significant and illogical variations of *CLW, identified and approachable in PFC1 process*

   c. *Identifying annual weaknesses* (per levels of importance): For the processes that have significant and illogical variations of *CCLW, identified and approachable in PFC1 process*

6. *Developing control points in every process of PFC1 (annual % of target touch [% TT]: Annual MCI target, based on annual MCI mean targets accepted in the company; annual MCI targets based on annual CLW targets and annual CCLW targets) (see Table 4.3, row C3):* $2,868,750 (approx. 5% of B2–B3; it may grow throughout the year up to 6% of B2–B3, depending on the evolution of the annual external profit expected from sales) (see Figure 4.7); *basic cost policy: utilization 100%*

7. *Verification of current causes, inspection items:* Verification of the *annual MCI means impact on the annual MCI targets* per levels of importance (of MCI means level 1, PST; level 2, *kaizen*; and level 3, *kaikaku*), more precisely on annual CCLW targets and on annual CLW targets and, implicitly, on annual losses targets and annual waste targets for every process of PFC1 (see Formulas 1.8 through 1.10). They are input data for the *annual MCI performance management* (confirmation of the annual MCI means; step 6 of the MCPD system) and for the *daily MCI management* (step 7 of the MCPD system).

8. *Verification of current causes, inspection points:* Verification % *of means target touch (% MTT) on the annual MCI targets*, more precisely on annual CCLW targets and on annual CLW targets and implicitly on annual losses targets and annual waste targets for every process of PFC1. Potential corrections of the AMIB may be proposed for existing and new products to the CEO in the month of July of year "N + 1." They are the input data for *annual MCI performance management* (confirmation of the annual MCI means; step 6 of the MCPD system) and for the *daily MCI management* (step 7 of the MCPD system).

9. Providing the input data for the development of the annual MCI action plan (time planning of the annual MCI means level 2, kaizen, and level 3, kaikaku).

10. Timely forwarding of the AMIB for existing and new products to the CEO for approval and amendments, together with the master budget and publication of budgets per cost centers; timely publication of budgets per cost centers (after completion of the catchball process for setting manufacturing improvement budgets; see Figure 4.12).

In conclusion, the annual target profit (annual operating targets profit) meant stability to the amount of $10,443,050. This target is over the initially planned amount of $10 million. The difference was planned to absorb the potential delays of the annual MCI means or other external and/or internal modifications. At the same time, the MCPD steering committee is aware that it may still increase the annual MCI targets up to 6% (to the amount of $3,442,500) of the annual CLW and CCLW identified and approachable to cope with the target of $10 million for the year "N + 1." This increase may be proposed in the month of July on the AMIB correction.

The level of the annual operating targets profit in the forecast budget is transferred to the forecast balance sheet and is the annual promise of the management team to the shareholders.

AMIB may be achieved for every existing or new product. For the future volumes of new products, the simulation will be made throughout their future processes knowing the opportunities for the annual MCI goal from similar products that experience the same processes. At the same time, AMIB is made for every PFC to see the impact on the total annual target profit and the contribution of each PFC to it.

Therefore, the annual stake of the MCPD system for PFC1 is of uncovering hidden reserves of profitability in the amount of *$2,873,050* (annual MCI goal) and of $10,443,050 (annual targets profits) by harmonious transformation of every process of the manufacturing flow corresponding to PFC1 by the fulfillment of the annual MCI means (kaizen and kaikaku) scientifically, systematically, and systemically (Figure 4.13).

At the same time, the annual stake from the perspective of price competitiveness, based on the impact of MCPD at the level of the product, is presented in Table 4.4 and Figure 4.14.

The effects of fulfilling the annual MCI goal will be seen in the months to follow. For that, the *AMCIB* is developed.

4.2.1.3 AMCIB for Each PFC

After completion of the first phase and the second phase of a manufacturing improvement budget cycle and concurrently with the development of AMIB, the AMCIB is drafted (see Figure 4.11, Phase 3).

The main purpose of the AMCIB is to provide for the level of effectiveness and efficiency and quantifiable in cash, for the annual MCI means, by meeting the annual MCI targets chosen to satisfy the level of the annual MCI goal.

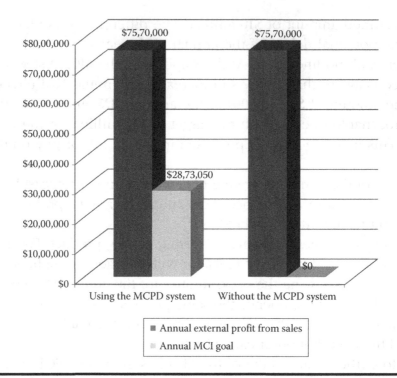

Figure 4.13 The stake of the MCPD system: manufacturing flow transformation for uncovering hidden reserves of profitability (annual MCI goal).

Table 4.4 Annual Operating Unit Profit and Unit COGS Analysis for a PFC1

	Using the MCPD System	*Without the MCPD System*
(A) Annual operating unit profit ($)	$12.29	$8.91
(B) Uncontrollable costs: fixed manufacturing budgeted expenses ($)	$0.59	$0.59
(C) Controllable costs: estimated manufacturing variable expenses ($)	$90.00	$90.00
(D) Controllable costs: annual MCI target (based on annual MCI mean targets accepted in company) ($)	$3.38	$0.00
(E) Sales and administrative expenses ($)	$0.51	$0.51
(F) Controllable costs: total annual CLW and CCLW identified and approachable for PFC1 in administrative process ($)	$0.01	$0.17
(G) Total cost of goods sold budget (COGS) ($) (B) + (C) + (E) − (D) − (F)	$87.71	$91.09

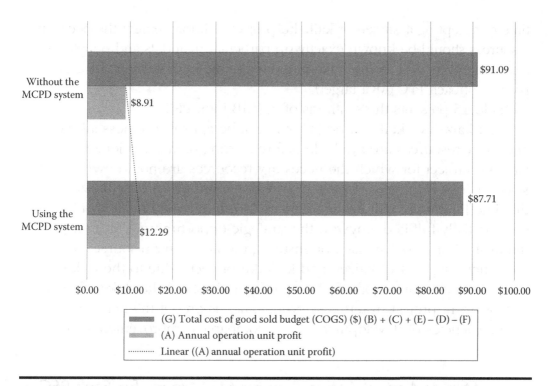

Figure 4.14   The annual stake of the MCPD system on unit profit for a PFC1.

So, AMCIB aims both at the correct (effectiveness) and the complete (efficiency) fulfillment of the annual MCI means to avoid and save CLW and CCLW at the level of each process of a PFC, and implicitly the reduction of the pressure on the capital budget at multiannual level. With the help of AMCIB, the foundations are laid for ensuring the self-financing capacity for the development of the company in the long run in order to avoid/reduce the loans outside the company.

AMCIB generates two cash flows (inputs and outputs). The timely fulfillment of the annual MCI targets may be done only by ensuring all the necessary resources, in particular for materials and people, especially that AMIB and AMCIB are continuously restoring themselves for the following 12 months (then for every term and month; it shall be corrected like AMIB twice a year—in July of the current year and in January of the next year).

So, as AMIB, AMCIB and setting the annual MCI targets and means are made almost in parallel to ensure an acceptable level of the annual MCI goal, and the person who will prepare AMCIB, from the MCPD steering committee, will continuously adjust the annual MCI means so that at the time of accepting a kaizen or kaikaku project of improvement the necessary resources should be known exactly (in particular materials and people, but also money), as well as the stake in money of the improvement for every process of each PFC (MCI target).

Table 4.5 presents the synthesis of AMCIB for a PFC.

So, a kaizen or kaikaku project will not be approved unless all the necessary resources are available before starting them. Any kaizen or kaikaku project for which the necessary resources are not known exactly, since their beginning, has great chances to fail (they will not fit within the time limit, they will never complete, or the planned results will be only partially fulfilled; anyway, the strategic opportunity of MCI target is diminished or lost). The lack of ensuring the necessary resources for the continuous support of kaizen and kaikaku projects, due to the lack of continuous determination of CLW to fully understand the stake of each improvement project in terms of profitability, is one of the main causes of the deficiencies of development of the companies' improvement culture.

4.2.2 Step 4: Annual Action Plan for MCI Means for Each PFC

The fourth step of the MCPD system refers to the development of the annual action plan for MCI means at the level of each PFC and further at the level of each department. Thus, every department and department manager develop annual and multiannual strategies for the purpose of meeting the annual MCI means targets and, implicitly, annual MCI targets and goal. These strategies are monitored by means of KPIs and their targets *driven by MCI means deployment* (product number increase, speed and agility increase, lead time decrease, productivity increase, product quality ratio increase), set based on annual reconciliation (see Sections 3.1.4.1 and 3.1.4.2).

At the same time, the annual MCI action plan, the fourth phase of manufacturing improvement budgets cycle (see Figure 4.11), is the part for budget execution of AMIB, by which every individual in the organization will know exactly and permanently his/her mission for the fulfillment of the annual MCI goal by participation in the annual MCI means (level 1, 2, or 3), more precisely by continuous reduction of CCLW and CLW in their work area. So, through annual MCI action plan, the aim is *total involvement of all departments and beyond and reduction or elimination of reactive managerial behavior.*

Table 4.5 AMCIB for Each PFC

| Annual MCI Targets and Means (Levels 2 and 3) for Each Process of Every PFCs (1…"n") | | Increase: Cash Saving with Kaizen to Meet Annual MCI Targets | | | Decrease: Cash Used for Annual MCI Means Level 2 (Kaizen) to Meet Annual MCI Targets | | | | | Decrease: Cash Used for Annual MCI Means Level 3 (Kaikaku) to Meet Annual MCI Targets | | Increase/Decrease in Cash | | |
| | | | | | Planned | | Actual | | | | | | | |
Cash and Equivalent Cash from Annual MCI Means for Each PFC to Meet Annual MCI Targets ($)		Planned	Actual	Achiev. (%)	Materials	Man * Hours	Materials	Man * Hours	Achiev. (%)	Cash for CAPEX	Payback (Years)	Planned	Actual	Achiev. (%)
Annual MCI targets for CCLW and CLW from equipment losses	Annual MCI means/process													
Annual MCI targets for CCLW and CLW from human work losses	Annual MCI means/process													
Annual MCI targets for CCLW and CLW from material/ energy losses	Annual MCI means/process													
Annual MCI targets for CCLW and CLW from waste	Annual MCI means/process													
Total ($)														

4.2.2.1 Annual List of Actions and Activities of MCI Means for Each PFC

Every department can influence certain processes along the manufacturing flow by daily activities that they develop. Every process is a bearer of CLW and CCLW. Through every CLW and CCLW, the annual MCI targets and annual MCI means are developed. In this context, every department will have an annual action plan to fight against annual MCI targets. Each kaizen and kaikaku project is directly connected to the annual MCI goal. For each kaizen and kaikaku project, interdepartmental teams are formed in order to meet the MCI means target.

In fact, every individual is continuously monitoring his/her area of activity in order to report from one shift to another the level of losses and waste and their connections throughout the whole manufacturing flow for each PFC to determine critical losses and waste. Based on such monitoring, CLW and CCLW are calculated for each process/work center/cost center, and targets are set in connection with annual MCI targets that are convergent with the annual MCI goal. In conclusion, every individual participates continuously, proactively, and preventively to setting the annual MCI targets for his/her area and he/she will also know how to meet the annual MCI means associated to the annual MCI targets because he/she participated both in setting the annual MCI targets and the annual MCI means by MCICP. This way, the pressure on setting the annual action plan at the level of every department is reduced and the continuous alignment to the annual MCI goal is enabled.

Figure 4.10 presents the allocation of every MCI target and MCI means target to a manager-in-charge. At the same time, at the level of one year, the contributions to fulfilling the MCI goal of the annual CCLW targets and of annual CLW targets will be determined (Table 4.6).

The allocated resources to meet the annual MCI means targets are monitored by means of AMCIB.

4.2.2.2 Annual Planning of MCI Means

All the annual kaizen and kaikaku improvement projects are monitored at the level of the department by the managers-in-charge, and then they are centralized by the MCPD steering committee (see Section 3.2.2.2). All the annual projects of strategic improvements that are convergent with the annual MCI goal are continuously monitored in order to prevent the reduction or elimination of any deviation that might stay in the way of fulfilling the annual MCI means (see Table 4.7).

Table 4.6 Annual Contribution of CCLW and CLW Targets to Meet the Annual MCI Means (Kaizen and Kaikaku) at Process Level of Each PFC

		Annual MCI Targets (from Current Status of CCLW and CLW to Target($))	Responsible Manager (Name and Surname—N/S; Contacts)	MS1 For Breakdown	MS2 For Setup	MS3 For Cycle Time	MS4 For Rework	MS5 For Scrap	MS6 For Motion	MS7 For Line Org.	MS8 For Energy	MS9 For Material	MS10 For Raw Material Stock	MS11 For Mistake Proofing	MS12 DOE/ANOVA	MS "n" Total 1 ($)	ms1 For Production Capacity	ms2 For New Processes Design	ms3 For New Technology	ms4 For New Product	ms5 For New Production Method	ms6 For New Equipment	ms7 For New Training System	ms8 For New Materials	ms "n" Total 2 ($)
Annual CCLW targets	Process—P1	XX%/$	N/S	X			X												X				X		
	Process—P2	XX%/$	N/S		X			X										X		X		X		X	
		XX%/$	N/S																						
	Process—P"n"	XX%/$	N/S																						
	Total ($) (1 + 2)			X	X		X									Total 1 ($)		X	X	X		X	X	X	Total 2 ($)
Annual CLW targets	Process—P1	XX%/$	N/S						X	X					X										
	Process—P2	XX%/$	N/S			X								X									X		
		XX%/$	N/S																						
	Process—P"n"	XX%/$	N/S																						
	Total ($) (1 + 2)					X			X	X				X	X	Total 1 ($)			X	X					Total 2 ($)

Table 4.7 Annual Planning of All Strategic Improvement Projects

I	II	III	IV	V	VI	VII
No. of Project for Annual MCI Means	Annual Type of Improvement	Annual Improving Code	Annual MCI Targets (%)	Percentage of Participation of Each MCI Target to the Annual MCI Goal	Annual CCIW and/or CLW Targets	Annual Losses and Waste Targets (%) (from Current Status to Target)
1	KAIZEN	MS1	%	%	%	TRL/PL %
2		MS2	%	%	%	TRL/PL %
3		MS3	%	%	%	TRL/PL %
"n"		MS "n"	%	%	%	TRL/PL %
1	KAIKAKU	ms1	%	%	%	TRL/PL %
2		ms2	%	%	%	TRL/PL %
3		ms3	%	%	%	TRL/PL %
"n"		ms "n"	%	%	%	TRL/PL %

VIII. Annual MCI Means—Defining/Preparing Improvements

- Type of Losses or Waste
- Code of Process/ Operation
- Responsible Department
- Responsible Manager
- Location for Improvement
- Defining Loss, Waste, and Costs
- Project Sponsor
- Leader of Project
- Improvement Team
- Workforce Training
- Type of Training: Internal/External
- Workshops
- On-the-Job Training

IX. Annual MCI Means—Solutions, Resources, and Benefits

- Solution for Improvement
- Possible Obstacles
- Material Resources Needed
- Man * Hours Needed
- Estimated Benefits
- Investment Needed
- Activities for Implementation
- Start Date
- Interim Data for Reporting
- End Date

X. Annual MCI Means—Evaluation Stage

- Stage 20%
- Stage 40%
- Stage 60%
- Stage 80%
- Stage 100%
- Actual Resources Consumed ($)
- Planned Resources Consumed ($)
- Public Presentations
- Delays/Review Deadlines
- Reasons for Delays
- Impact Delay in Costs

XI. Annual MCI Means—Gains

- Time
- Production Capacity
- Distances
- Motion
- Direct and Indirect Labor
- Stocks: Materials, WIP, Goods
- Intangible Effects
- Investment Avoided

XII. Annual MCI Means—Project Evaluation

- Benefits vs. MCI Targets
- MCI Targets vs. Actual Costs
- Benefits vs. Time
- Team Evaluation—Gains
- Team Leader Evaluation—Gains
- Total Actual Benefits ($)
- Total Planned Benefits ($)
- CAPEX Spent on MCI Projects
- Benefits/Costs for 12 Months (%)
- Payback: MCI Projects (Years)
- Solution Validity over Time

As in Table 4.7, each MCI target of a process of each PFC (column IV) has a level of participation in fulfilling the annual MCI goal (column V) by fulfilling the annual CCLW targets and annual CLW targets (column VI), implicitly of annual losses and waste targets (column VII), to attain the annual MCI means target (columns VIII, IX, X, XI, and XII). Thus, *all the people in the company and beyond are aligned continuously for the fulfillment of the MCI goal.*

4.2.2.3 Definition and Preparation of Individual Plans for MCI

In this moment, all annual MCI targets and means are already set to fulfill the annual MCI goal and every manager knows exactly what he/she is responsible for in a MCPD system for the following year. It is time to present to all the employees their assignments to meet the annual MCI means targets in the processes where they are involved: Who? What? When? How? How well? With whom? and Why? (see Section 4.1.2.6).

Table 4.8 presents an example of presentation of the individual annual MCI means for a kaizen improvement project for setup, using SMED and MDC technique (elimination, combination, rearrangement, and simplification) (Posteucă and Sakamoto, 2017, p. 332). As we can see, the kaizen improvement project for setup has a level of importance 4 out of 15 at the level of year "N + 1" from the perspective of the annual MCI goal.

Internal communication of the individual annual MCI means refers to the presentation of the connections between the stake in money of the improvement project (CCLW targets for setup for process "Z"—KKIs) and OMIs (in our case, company lead time and especially the annual MCI goal).

4.3 Manufacturing Cost Policy Management

The purpose of manufacturing cost policy management, the third and last phase of the MCPD system, is to

- Continuously attract workforce to fulfill the MCI goal, by meeting the annual MCI targets and means, based on an annual MCI action plan later developed, by the internal organization of the company in this

Table 4.8   Individual Annual MCI Means

"Plant A"		"Plant A"		Production Department		Interdepartmental		Individual DMIs		Individual Activities		Progress				Comments
OMIs	Target: Year: "N+1"	KPIs	Target: Year: "N+1"	KPIs	Target: Year: "N+1"	KKIs	Target: Year: "N+1"	DMIs	Target: Year: "N+1"	Specific Activity	By when	Q1	Q2	Q3	Q4	
Company lead time Annual MCI goal XXXX$	−7%	Production lead time for "X" PFC	−8.5%	Annual MCI target (degree of importance 4 from 15)—CCLW from setup for process "Z"	9 min <	Setup time for "Y" equipment (injection molding machine—vertical mold change) CCLW targets for setup for process "Z"	9 min <	Number of training/workshop hours for setup	6 training hours/operator (in 2 days)	Participation in training setup reduction (SMED and MDC) (including viewing and studying the movie archive for setup (group intranet)	01.06 "N+1"			X		During limit
										The work done						
									18 hours/operator in kaizen for setup	Participation in kaizen for setup reduction: data collection, analysis equipment operating principles, etc.	01.09 "N+1"			X		During limit

respect and by training programs and workshop corresponding to MCI (*Step 5: Engage the Workforce to Achieve the Annual MCI Targets*).

■ Continuously monitor the level of performance concerning the fulfillment of the annual MCI targets for the fulfillment of the annual MCI goal or, more precisely, to continuously monitor the level of meeting the annual MCI means targets (root causes, losses and waste, and CLW and CCLW), which, together with the level of variation of manufacturing costs, determine the extent of fulfillment of the annual MCI targets (effects, expected results), Phase 5 from Figure 4.11 (*Step 6: MCI Performance Management*).

■ Ensure the development of daily MCI management at the shop floor level by monitoring tangible KPIs of MCI (of losses and waste and of CCLW and CLW) and their connections with other KPIs, the continuous development of management branding (MB, *contextual managerial behavioral identity*) (Posteucă, 2011; Posteucă and Sakamoto, 2017, pp. 240–244) and continuous development of daily management for the purpose of ensuring a better operational performance of MCI and fulfillment of MCI means (in particular level 1, PST, and level 2, kaizen) (*Step 7: Daily Manufacturing Cost Improvement Management*).

So, by this last phase of the MCPD system, it is aimed at fulfilling the *total and continuous involvement of all departments and beyond*, the reduction and/or elimination of *reactive managerial behavior*, and, last but not least, the reduction and/or elimination of *incorrect and incomplete improvement projects implementation*.

4.3.1 Step 5: Engage the Workforce to Achieve the Annual MCI Targets

In order to ensure the total attraction of people to the MCPD system, in order to meet the annual MCI targets for the purpose of fulfilling the annual MCI goal, a department organization and continuous learning is required to support the pro-cost and pro-productivity culture ("Do" phase from the PDCA cycle).

The way the company is organized to achieve the production for the purpose of obtaining an acceptable level of annual and multiannual external profit from sales, a similar organization of the company to meet the multiannual internal profit and the annual MCI goal is required. Every individual in the company must know his/her annual role in accomplishing the operational assignments for the fulfillment of annual external profit from sales/production (organization structure of the company, e.g., matrix organization structure and the annual learning plan—technical training, workshops, and on-the-job training/direct instruction) and to fulfill the assignments corresponding to the annual MCI goal (organization for MCI [see Section 3.2.2] and annual training plan for the timely meeting of the annual MCI means targets and of the targets for SKPMP [see Figure 4.1]).

4.3.1.1 Departmental Organization for Achieving the Annual MCI Targets

In order to organize the company for meeting the annual MCI targets and means, it is necessary to allocate assignments to every manager in this respect (see Section 3.1.6), besides the current operational tasks for the achievement of the production plan. Every department manager will have throughout the year a number of improvement projects (MCI means) that he/she will coordinate and for which he/she will form interdepartmental teams. Moreover, the department managers, in order to make sure that the annual MCI means targets will be met in time, will develop certain annual and multiannual activities specific for each department; they will develop a series of departmental KPIs to monitor the evolution of CCLW and CLW for every process that they can, to set targets for KPIs developed for CCLW and CLW, and will develop annual and multiannual plan for departmental activities to meet MCI means targets.

Table 4.9 presents the way a manufacturing company is organized to accomplish the annual MCI action plan to meet the annual MCI goal (column V).

As we can see, every department is responsible for an annual number of kaizen and/or kaikaku projects (column II) to meet the annual MCI targets (column IV) by annual CCLW and/or CLW targets (column VI) and by annual losses and waste targets (%) (from current status to target) (column VII). At the same time, every department manager (MCI group leader) will be responsible of a certain annual number of kaizen and/or kaikaku projects of the processes they can influence the most (Θ, strong relation) or will allocate people in other projects in other departments (Q, some relations; Ŏ, limited relations).

Table 4.9 Engage the Departments to Achieving the Annual MCI Targets

I — No. of Projects for Annual MCI Means (Levels 2, Kaizen, and 3, Kaikaku)	II — Type of Improvements (Means)	III — Annual Improving Code	IV — Annual MCI Targets (%)	V — Percentage of Participation of Each MCI Target to the Annual MCI Goal	VI — Annual CCLW and/or CLW Targets	VII — Annual Losses and Waste Targets (%) (from Current Status to Target)	VIII Factory Departments — Production	Maintenance	Quality	Industrial Engineering	Internal Logistics	External Logistics	Research and Development	Environmental, Health, and Safety	HR	Finance and Accounting
1	Kaizen	MS1	%	%	%	TRL/PL %	Θ	Ò	Θ	Ò	Q	Ò	Ò		Q	
2		MS2	%	%	%	TRL/PL %		Ò	Q			Q		Θ		
3		MS3	%	%	%	TRL/PL %		Θ			Ò	Ò		Ò		Q
"n"		MS "n"	%	%	%	TRL/PL %	Ò	Ò	Ò	Ò			Q		Θ	
1	Kaikaku	ms1	%	%	%	TRL/PL %		Ò		Q	Θ	Θ		Ò		
2		ms2	%	%	%	TRL/PL %	Q	Q	Ò		Q	Θ	Ò		Ò	Θ
3		ms3	%	%	%	TRL/PL %					Ò		Θ	Q		Ò
"n"		ms "n"	%	%	%	TRL/PL %			Ò	Θ		Ò		Ò		Ò

Notes: Θ, strong relation; Q, some relations; Ò, limited relations; MS, means for systematic improvement; ms, means for systemic improvement.

4.3.1.2 Sources for Determining Training Needs Related to the Fulfillment of the Annual MCI Targets

From the perspective of the MCPD system, any training program and workshop must approach a concrete problem of the company (operational, external profit, and/or MCI, internal profit). The annual cycle of development of the training needs follows the PDCA cycle, that is

1. *Determining the training needs and the associated budget* (budgets that are embedded in AMIB—decrease cash used for the annual MCI means to meet the annual MCI targets)
2. *Determining the annual training plan* (for the operational tasks/fulfillment of the production plan and for the tasks required to meet the annual MCI means targets)
3. *Selection of annual training methodologies* (theoretical training, workshops, KPIs monitoring, on-the-job training, best practice study in group or other departments, OPL)
4. *Establishing the annual training assessment system* (written tests and/or practical tests in areas specially designed for simulations of the theory already learned)
5. *Developing the annual calendar of training programs to support the expected results* (annual MCI means targets)
6. *Continuous measurement of the effects of training programs at the level of processes of each PFC* (monitoring the achievement of operational KPI targets and annual MCI targets; the results of the training programs must be visible and measurable) (Posteucă and Sakamoto, 2017, pp. 119–123)
7. *Developing training plans for future periods and plans to extend positive and negative experiences during the annual training programs in other areas of the company or other companies in the group*

Every individual in the company should participate at least once every two years in an internal or external training program to support the annual MCI means targets.

In order to support the continuous development of the people's skills and knowledge and implicitly of a company where people are continuously learning, for the purpose of supporting the current and future transformation of the manufacturing flow and customer satisfaction, it is necessary to identify and set the priorities of the annual training program (*annual training needs*).

Therefore, annual training needs will aim at both the support of current tasks of people (SOP) and the MCI (SOP improvement from the perspective of MCI). The source determination of annual training needs for the current or future operations is as follows:

1. Monitoring the level of performance of regular activities by the team leader in order to propose training specific for each process/zone in the company
2. Monitoring the level of performance of the additional activities by the workplace responsible to propose training specific for additional activities for each process/zone in the company
3. Monitoring the level of performance of regular and additional activities by specialists, auditors, and operators to propose training programs and OPL
4. Analysis of the need for an increase of flexibility and versatility of operators for every change and/or equipment or line, by workplace responsible to propose training programs and on-the-job trainings
5. Analysis of the needs of skills development identified by every employee and by workplace responsible to propose training programs

The source determination of annual training needs to meet the annual MCI means targets and implicitly of annual MCI targets for each PFC is as follows:

1. Continuous monitoring of the level of performance of losses and waste (Posteucă and Sakamoto, 2017, pp. 119–123) and of CLW and CCLW by the employees, team leaders, and responsible workplace, with the help of specialists in cost management and of members of the MCPD steering committee, to propose training programs and workshops to learn the methods as continuous improvement instruments of manufacturing cost
2. Analysis of future technologies, equipment, and methods of management by industrial engineers and by department managers in the workplace responsible to propose training programs and workshops to reduce CLW and CCLW generated by the lack of knowledge on current work with these new technologies, equipment, and methods of management (prevention of CLW and CCLW)

Identification of the annual needs for training and training programs must support the meeting of the annual MCI targets and of MCI "destinations" of each PFC (see Figure 2.7).

4.3.1.3 Initial and Updated Annual Training Plan for Operators, Supervisors, and Managers to Achieve the Annual MCI Targets

Every year it is necessary to develop a systematic plan for the training programs to support the MCI means targets with a view to increase the skills and knowledge of operators, supervisors, specialists, and managers.

The examples of competences aimed to be acquired are as follows: enhancement of the planning, supervising, and motivating skills of the teams implementing the MCPD system; increased accuracy of setting the CLW and CCLW; the method of setting MCI targets and means; improvement of delegation skills; refinement of the skills in monitoring the CCLW and CLW; advancement of the skills in choosing the methodologies for improving the processes; increase in knowledge and understanding of the business; enhancement of skills to approach and solve a problem; enhancement of skills to make improvements, kaizen; refinement of the skills to collect losses and waste; etc.

Table 4.10 presents an example of an MCI training plan. As we can see, every training program is connected to the annual MCI goal (column V) and implicitly annual MCI targets (column IV).

The annual MCI training plan is adjusted in the month of July of the current year and in January of the next year (updated annual MCI training plan), together with the annual MCI targets and means.

4.3.1.4 Running the Activities and Actions of the Annual MCI Means to Meet the Annual MCI Targets

After the thorough planning of the annual MCI means, the next phase is the effective work to meet the MCI means targets. All the people involved should be aware that the effects of the improvement they are working on will help them better perform their routine tasks, even if it involves changing the SOP. Members of the improvement teams, especially kaizen projects, will receive from the *MCPD steering committee* the CLW or CCLW description and their implications in the company's processes and need for development, based on concrete data and information (charts, trends,

Table 4.10 Annual Training Plan for Operators, Supervisors, and Managers to Achieve the Annual MCI Targets

I The Number of the MCI Training Program	II Type of Improvements (MCI Means)	III Annual Improving Code	IV Annual MCI Targets (%)	V Percentage of Participation of Each MCI Target to the Annual MCI Goal	VI Annual CCLW and/or CLW Targets	VII Annual Losses and Waste Targets (%) (from Current Status to Target)	VIII–XV	XVI Type of Participant: Operators (O), Supervisors (Su), Specialists (Sp), Managers (M)	XVII Production	Maintenance	Quality	Industrial Engineering	Internal Logistics	External Logistics	Research and Development	Environmental, Health, and Safety	HR	Finance and Accounting	XVIII	XIX
1	Kaizen	MS1	%	%	%	TRL/PL %		O, Su, Sp	Θ	Q	Ǒ	Q		Ǒ			Q			
2	Kaizen	MS2	%	%	%	TRL/PL %		O, Su, Sp,		Θ	Θ		Q			Q		Q		
3	Kaizen	MS3	%	%	%	TRL/PL %		O, Su, Sp, M		Ǒ				Q		Ǒ				
"n"	Kaizen	MS "n"																		
1	Kaikaku	ms1	%	%	%	TRL/PL %		Su, Sp, M			Θ		Q		Q			Θ		
2	Kaikaku	ms2	%	%	%	TRL/PL %		Sp, M		Θ			Q		Q					
3	Kaikaku	ms3	%	%	%	TRL/PL %		Sp, M				Ǒ	Q				Ǒ			
"n"	Kaikaku	ms "n"																		

Column headings (VIII–XV):
- VIII — Beneficiary of the Training: Department/Process/PFC
- IX — The Name of the Training Program
- X — Description of the Practical and Theoretical Need of the Training Program (Objectives to Be Achieved)
- XI — Month(s) and Day(s) for the Training Program
- XII — Hours of Training and Workshop
- XIII — Type of Training: Internal/External
- XIV — Lecturer Name/Training Company
- XV — List of Training Participants (Name and Surname and Contact Details of Each)

XVII — Plant Departments Involved in MCI's Annual Training Programs

XVIII — Establishing the Training Evaluation System

XIX — Measurement of the Effects of Training Programs at the Level of Processes of Each PFC/Department (Immediately after Training /Continuous/Over Time)

Notes: Θ, strong relation; Q, some relations; Ǒ, limited relations; MS, means for systematic improvement; ms, means for systemic improvement.

positioning on layout, etc.). Then the interdepartmental team will initially analyze the MCI means target and set the main objectives (especially that the improvement project must be completed in a maximum of three months, kaizen, and a maximum of six months, kaikaku) and the data required to be collected by each member (what parameters should be improved, until when, and what are the current CLW and CCLW—measured by the team). Then the teams analyze the phenomena and principles and broadly define the problems and causes of MCI and propose solutions for improving CLW and/or CCLW. These solutions must satisfy the MCI means targets level and, implicitly, the MCI target. The necessary resources are assigned for the solutions proposed to achieve the MCI means targets and they proceed to their implementation. Subsequently, MCI performance targets are monitored by the MCI performance management.

4.3.2 Step 6: MCI Performance Management

Annual MCI performance management refers to the effectiveness and efficiency of the annual MCI means at the level of each PFC process to meet the annual MCI targets and, implicitly, the annual MCI goal, namely, the AMIB and AMCIB performance. At the same time, the annual MCI performance management refers to the degree of involvement of all employees in the company in achieving the annual MCI means to ensure the company's ongoing transformation in the direction imposed by the market. The transformation efforts of a company start from the multiannual target profit and end with the multiannual target profit. The results of MCI performance management must be visible at the level of each PFC process, that is, directly, the continuous and planned reduction of unit manufacturing costs to meet the annual MCI goal and, indirectly, increase in productivity and quality in each manufacturing process. In fact, MCI performance management verifies the confirmation of the results regarding the annual MCI goal completion level (PDCA cycle check phase).

4.3.2.1 Performance Evaluation of AMIB for Each PFC

The performance evaluation of AMIB refers to the degree of achievement of the annual MCI goal (see phase 5, Figure 4.11). The evaluation is performed twice a year in July of the current year and January of the next year.

As shown in the previous AMIB example (see Section 4.1.1.2) and in the formulas in Chapter 1 (see Formulas 1.2 through 1.6 and 1.8 through 1.10), the annual MCI goal variation analysis is performed for each process of each PFC. The purpose of the analysis is to identify the seven levels of variations for each process of each PFC. These variations will seek to identify the reasons why the annual MCI targets at the level of a process were not achieved and to propose the annual MCI means targets rectifications, so that at the end of the year both negative variations of the annual MCI targets and potential negative variations of the annual external profit expected from sales were not achieved, to meet the annual target profit (achievement of *Principle No. 1 of the MCPD System: Target Profit from MCI Does Not Change*). Negative variations mean those variations that contribute to the failure to meet the annual MCI goal/annual MCI targets.

Therefore, the purpose of the annual MCI goal variation analysis is to identify new directions for reaching the annual MCI targets and not to just look for the reasons why the targets were not achieved. Excuses within the MCPD system are futile. In this context, planning and replanning of the annual MCI goal represents the true value of a managerial team within a company. Figure 4.15 presents the annual MCI goal variation analysis model.

AMCIB performance evaluation refers to the effectiveness and efficiency of the annual MCI means. Effectiveness means the level of achievement of the annual MCI means targets, and efficiency means the annual MCI means targets falling in the target level of resources consumed. In this context, a kaizen project (MCI means level 2) can be effective, fit into the established performance targets; more exactly it met the MCI targets but may not have been efficient because it consumed more resources than planned. Conversely, a kaizen project may be efficient and may not be effective. Any MCI means must be both effective and efficient to be performed.

4.3.2.2 Assessment of the Degree of Employee Involvement to Achieve the Annual MCI Targets

Ensuring an organization to facilitate the participation in the annual MCI means (see Section 4.3.1.1), developing the annual MCI training plan (see Section 4.3.1.3), establishing the annual MCI means (see Table 4.8), and developing the AMCIB to ensure timely all the necessary material and human resources create the premises for high employee participation in the annual MCI means and the provision of a MCI benchmark.

However, there may be a number of objective and especially subjective factors that need to be monitored and addressed by analyzing the desired

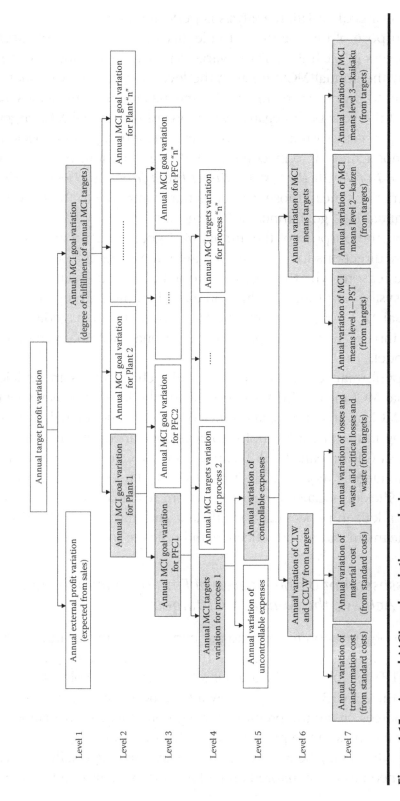

Figure 4.15 Annual MCI goal variation analysis.

state and the current state of the involvement of all people in and beyond the company to achieve the annual MCI targets.

These are the following factors:

1. The level of internalization of the MCPD system concepts by managers
2. Continuous support of managers in reaching the annual MCI targets and means
3. The degree of interest of new employees for their actual participation in meeting the annual MCI means targets
4. Documenting each process from the annual CLW and CCLW targets perspective
5. The depth and quality of lifelong learning programs and the evaluation of the results obtained over time
6. Omitting negative results in the annual MCI targets and means reports
7. Limiting people's participation in achieving the annual MCI means by planning a large volume of production, diversified and in small batches
8. Changing production plans during a month, a quarter, or a year
9. Launching more new products than planned, especially in short time intervals
10. The lack of real collaboration with the MCPD consultant

Therefore, planning all MCPD system activities beyond production hours and thorough preparation of all MCI means-related activities is needed to continuously capture the attention of all people in and beyond the company, people who have tasks on meeting the annual MCI goal.

In this context, the annual assessment of the extent to which people are involved in achieving the annual MCI means targets is performed through the planning and continuous evaluation of the following elements (by the members of the MCPD steering committee):

■ The number of improvement ideas per year/person/areas/processes/ PFC/plant (through the internal suggestion system)
■ The number of kaizen projects deployed per year/person/areas/ processes/PFC/plant
■ The number of solutions implemented as a result of kaizen projects (SOP creation or adjustment) per year/areas/processes/PFC/plant
■ The number of certifications/qualifications per year/person/areas/ processes/PFC/plant

- The number of interim and final public presentations of kaizen and/or kaikaku project outcomes per year/areas/processes/PFC/plant
- The number of training per year/person/areas/processes/PFC/plant
- The number of workshops per year/person/areas/processes/PFC/plant
- The number of kaizen/kaikaku projects delayed

The continuous improvement of the involvement of people in the company and beyond creates the premises for employee morale growth, the achievement of the annual MCI goal and annual external profit expected from sales/production. The fulfillment of the annual MCI means targets must be visible to increase employee confidence in the MCPD system and to reduce stress levels, especially among middle managers.

4.3.2.3 Systemic Cost Improvement Performance Management

The systemic approach to cost improvement involves understanding the phenomena and principles of manifestation of productivity and quality issues at the level of each PFC process. From the perspective of the MCPD system, overcoming the standard level of stock quantities, especially WIP, caused by the change in the current loss standard level (CAL), at which the production volume was planned, is called waste elasticity of losses. These variations are called *waste (stocks) elasticity on the losses* (Posteucă and Sakamoto, 2017, pp. 236–238).

For example, the coefficient of waste elasticity (stocks) depending on OEE is determined by comparing the percentage modification of waste (stocks), to the percentage modification of CAL for OEE.

$$E_{woee} = \frac{\Delta W\%}{\Delta OEE\%} = \frac{\Delta W/W_0}{\Delta OEE/OEE_0} = \frac{\Delta W}{\Delta OEE} \cdot \frac{OEE_0}{W_0} \qquad (4.1)$$

where
  $\Delta W\%$ is the percentage modification of waste (stocks) between the initial planning and current status
  $\Delta OEE\%$ is the percentage modification of OEE between the standard level and the current level

Similar to the formula for (*Ewoee*), *human work elasticity on OEE (Ehwoee)* and *materials and energy elasticity on OEE (Emeoee)* are determined.

Moving forward, CCLW approach to the entire manufacturing flow of a PFC involves understanding the effects of variation of current standards of cycle time of the bottleneck, process capacities, setup, and preset transfer time between processes of a critical process over the other processes of a PFC or more PFCs (see Formula 3.9). The effects of these variations are found in the increase of the CLW level related to upstream and downstream processes considered critical.

Therefore, the systemic approach of MCI requires a holistic approach, by fully understanding the *waste (stocks) elasticity on the losses* in a critical process that generates *CCLW* along the entire manufacturing flow of a PFC (or over the processes of several PFC). The systemic approach of MCI involves planning the standardization and stabilization efforts of each process through kaizen activities (MCI mean level 2).

In conclusion, MCI performance management involves achieving the annual MCI goal by meeting the annual MCI targets and means.

4.3.3 Step 7: Daily Manufacturing Cost Improvement Management

Daily MCI management refers to daily processes control from the perspective of nonstrategic changes at MCI level. The annual MCI goal control cycle (through the PDCA cycle), implicitly in the MCPD system, involves establishing the annual MCI targets and means, developing AMIB and AMCIB, developing annual MCI action plans, engaging the workforce to achieve the MCI targets, checking the results through MCI performance management, and incorporating results into MCI targets and means and action plan for the next period. The key point of achieving the annual MCI goal is identifying the critical items (annual control points of MCI) that require improvements. Specifically, the PDCA cycle from the annual MCI goal perspective assumes the following:

1. Establishing and deploying the annual MCI targets and means at the level of each process of each PFC (*plan*)
2. Implementing the annual MCI means and solving the annual weaknesses of MCI (*do*)
3. Performance evaluation and verification of progress toward annual MCI targets (*check*)
4. Standardization of results in daily controls or the restoration of the annual MCI action plan (*act*)

Therefore, daily MCI management involves defining and conducting all necessary activities at the level of each department and at the level of each PFC process so that MCI targets are kept under control by a daily effective and efficient controls ("Act" phase from the PDCA cycle).

4.3.3.1 Daily Monitoring of Tangible KPI of MCI

Daily monitoring of MCI involves checking the current state of the annual MCI targets (effects or results) and the annual MCI means (cause or factors) to react to any significant deviation that affects the annual MCI goal. Implementing solutions for MCI improvement projects involves establishing a standard. Improvement targets are set up on a continuous basis for this standard (MCI means level 2—kaizen). At the same time, this standard is controlled daily to make sure its current level is under control. When a standard can no longer meet the market requirements (production opacity, quality level, and/or cost level), it is replaced and a systemic improvement (MCI means level 3—kaikaku) is implemented.

Therefore, the daily control of current MCI standards until they reach the next level, improved by a kaizen project, represents the goal of daily MCI management.

In this context, tangible KPIs of daily MCI management for each process of each PFC are as follows:

A. *Checking current results with MCI targets* (for each process of each PFC; effects or results):
   a. *Control item* (elements that contribute to meeting the annual MCI targets):
      1. KPIs of transformation cost (variable costs) (see Section 3.1.2.2)
      2. KPIs of material costs (variable costs) (see Section 3.1.2.2)
      3. KPIs of CLW (see Sections 3.1.3.3 and 4.1.1.3)
      4. KPIs of CCLW (see Section 3.1.3.3)
   b. *Control point* (checking the annual weaknesses from the processes of a PFC used to plan the achievement of the annual MCI targets for a PFC; checking the *annual % of MCI target touch (% TT)) (see Section 4.2.1.2)*:
      1. KPIs of the annual weaknesses of *manufacturing controllable expenses*, identified and approachable on each PFC process (on importance levels; control of actual level)

2. KPIs of the annual weaknesses of *CLW*, identified and approachable on each PFC process (on importance levels; control of actual level)

3. KPIs of annual weaknesses of *CCLW*, identified and approachable on each PFC process (on importance levels; control of actual level)

B. *Checking current results with MCI means targets* (for the annual MCI means levels 1, 2, and 3 for each process of each PFC; checking the current state of causes/factors that contribute to meeting the annual MCI targets):

a. Inspection item (activities and actions contributing to the achievement of the annual MCI means targets):

1. Impact of MCI means targets level 1, PST, on the annual MCI targets (or on the annual CLW targets and annual CCLW targets; keeping the current standard of MCI; problem-solving task)

2. Impact of MCI means targets level 2, *kaizen*, on the annual MCI targets (or on the annual CLW targets and annual CCLW targets; systematic improvement task)

3. Impact of MCI means targets level 3, *kaikaku*, on the annual MCI targets (or on the annual CLW targets and annual CCLW targets; systemic improvement task)

b. Inspection point (checking the current state of the achievement level of % *MTT on the annual MCI targets*):

1. KPIs for losses, process level (Posteucă and Sakamoto, 2017, pp. 119–123) (see Figure 4.7, matrix 2)

2. KPIs for waste, process level (Posteucă and Sakamoto, 2017, pp. 119–123) (see Figure 4.7, matrix 2)

3. KPIs for transformation cost and material costs, process level (see Section 3.1.2.2)

The establishment of the control item at the level of each PFC process is performed concurrently with establishing the annual MCI targets and means (annual MCI policy). Through daily MCI management, every person in the company knows exactly the level of meeting the MCI targets and means for his/her area. Information is updated at regular intervals. The information is verified in conjunction with one or the other. The major challenge is to know the manufacturing cost level for each process as quickly as possible to determine CLW and then CCLW, especially manufacturing overhead costs.

4.3.3.2 Daily Management: KPIs of MCI and Other KPI

Every job can have two types of tasks: (1) routine tasks (to meet the annual external profit expected from sales/production) and (2) tasks to improve and innovate the workplace (to meet the annual MCI goal, and also the annual external profit).

If all people in the company and beyond would be involved almost exclusively in performing routine tasks at the level of a process or the entire manufacturing flow of a PFC, even if the quality level is very good and the work speed is good, the reduction of unit manufacturing cost by itself cannot be achieved and then the annual MCI goal cannot be met. Routine activities cannot cover the solving of a quality or productivity problem and, at the same time, the achievement of the MCI targets. The three directions of people's work (cost, timing, and quality) are often antagonistic. Companies often choose to act mainly in one of the three directions. Through the MCPD system, the three directions are aligned in order to achieve the annual MCI goal and the annual target profit. For this reason, tasks to improve and innovate the workplace become as important as routine tasks. In fact, the two types of tasks are like two sides of the same coin: annual targets profit. Concerns about designing tasks to improve and innovate work should be the focus of senior and middle managers.

Monitoring people's work and verifying the level of reaching the KPIs targets for the two types of tasks are performed at the same time as they are inseparable.

Figure 4.16 presents the symbiosis between KPIs related to routine tasks (*safety, current cost, quality, production,* and *delivery*) and MCI-related KPIs (inspection items/points and control items/points).

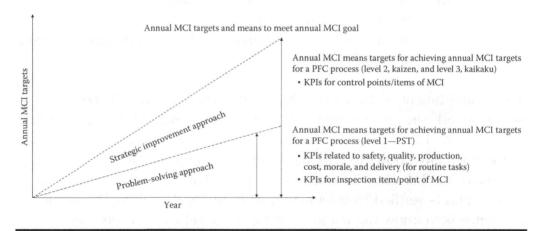

Figure 4.16 Daily management: symbiosis between KPIs for routine tasks and KPIs for MCI.

The response to deviations from the acceptable level of losses and waste must be immediate. One does not have to wait until the balancing of account of the month to react. Clear responsibilities and alert rules are established at each hierarchical level in order to provide conclusions for each type of MCI problem at the shop floor level.

For continuous monitoring of these KPIs, the MCI PFC Status Board, the MCI Department Status Board, and the MCI Process Status Board are used.

4.3.3.3 Management Branding for MCI

Discussions about company profits and about costs between managers and employees are and will remain sensible. Meeting MCI targets and means for PFC processes depends on managers' responses, prompt responses, and their expected behavioral identity. Often managers' willingness to listen to an employee on an MCI problem is diluted as the first MCPD system implementation period (first 12–18 months) passes.

The managers' dilemma is: *how long do I have to allocate from my program for the external profit/production and how much for the annual MCI goal/improvements?* Some managers remember the need for improvement only when manufacturing is within the set limits or when an external audit of improvements is approaching. Otherwise, they are almost exclusively concerned with the production. If the main goal of any improvement is to stabilize a process, through standardization, and then continuously improve that standard to meet the current and future process needs (obtaining the external profit), in fact to continuously ensure change toward the better (transformation), then *managers need to spend more time on the strategic improvement approach than on the operational activities* by continuously assuming the design and redesign of the manufacturing flow in order to achieve the productivity mission (ensuring long-term production capacity and profitability plan).

From the perspective of the MCPD system, the message about the need to meet the annual MCI goal must be constant, as does the need to achieve the external profit. Real engagement and managers' behaviors are extremely important from the point of view of ensuring a continuous transformation of the manufacturing flow to achieve the multiannual and annual profit targets. The continuous monitoring and development of a managerial behavior that is consistent with the expectations of employees, especially in crisis situations, creates the premises of a company focused on learning, creativity, and innovation on a continuous basis. This expected management behavior is called *management branding.*

Management branding "is a managerial system that, by an integrated approach, creates and synchronizes, for the application, contextual managerial behavioral identities in order to increase organizational productivity and/or economic growth" (Posteucă, 2011).

Therefore, the identity of the managers must be pro-productive and pro-cost. This identity must be perceptible by every person in the company and beyond. Continuous learning, innovation, and creativity for people need to be convergent with the continuous improvement of lead time and with the continuous reduction of the manufacturing cost, especially the transformation cost.

Figure 4.17 shows the MB cycle following the PDCA steps. It starts from analyzing the undesirable behavior of managers and choosing the behavior to be improved (plan) to continuously capture the differences between the necessary behaviors and the actual behaviors of managers at different levels of the company (Posteucă and Sakamoto, 2017, pp. 240–244). Managers

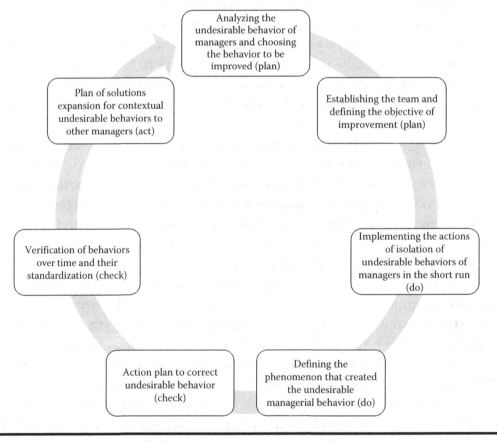

Figure 4.17 Management branding: continuous improvement of contextual managerial behavioral identities.

address their behavioral problems horizontally, that is, the issues are dealt with peer to peer, on the same hierarchical level and, in exceptional cases, escalation is performed to a higher level. This approach enhances employee morale and involvement by encouraging and implementing the managerial behaviors expected by others and facilitating the implementation of ideas and suggestions from shop floor colleagues.

Therefore, the manager of the MCPD system should be perceived as a coach and a mentor to be continually open to improvement and innovation.

4.3.3.4 Daily MCI Management Process

The continued promotion of the MCPD system implies the development of an atmosphere that emphasizes the growth of the initiatives of all people.

The daily MCI management process is a set of principles, processes, and tools that allow monitoring and obtaining the annual MCI targets and means at all levels of the organization and clear setting of the decision-making levels by designating owners for all MCI-related KPIs. The daily MCI management process ensures fast horizontal and vertical communication to continually react quickly and respond appropriately to any problem with the achievement of the annual MCI means targets.

The daily MCI management process involves the use of a visual management that continually shows the current level of performance of the annual MCI targets and means for each PFC process through short-term control of MCI inspection points. The time allocated for the daily physical presence of the daily MCI process management is as follows: (1) plant manager, 10%–20%; (2) production manager, 20%–40%; (3) department managers, 50%–60%; (4) MCPD managers, 50%–60%; and (5) team leaders, 75%–100%.

The daily MCI management process ensures quick and right response to MCI targets deviations (in shop floor KPIs) with short control intervals. The information must be immediately visible at the level of the deviation process. At the same time, there must be a clear responsibility for MCI results at all levels (MCI control items/points).

In this context, the daily MCI management process consists of the following steps:

1. *MCI performance monitoring*: For MCI control items/points and MCI inspection items/points.
2. *Problem reaction*: For MCI inspection point and initiation of MCI performance meeting analyses.

3. *Prioritization of problems*: For MCI inspection point (daily top five deviations of MCI).
4. *Problem solving*: 5 *Whys?* or *A3* to solve the deviations problems from MCI targets (MCI control points).
5. *Check implemented solutions*: Checking the consistency over time of the solutions chosen to solve MCI problems.
6. *Plan to extend the solutions*: Replicating solutions that have had good results for other similar processes.
7. *Future plans for systematic MCI*: These plans help moving from MCI means level 1, PST, to MCI means level 2, kaizen. Specifically, what could not be addressed through PST will be tackled through kaizen; the annual MCI targets must be met.

Consequently, the daily MCI management is, through a horizontal and vertical communication system, a process of identifying deviations from the MCI inspection point, collecting data and information about causes, analyzing root causes, choosing the most appropriate solutions for that time, and monitoring the consistency over time of the solutions implemented.

References

Akao, Y., 1991. *Hoshin Kanri: Policy Deployment for Successful TQM* (originally published as Hoshin Kanri Kaysuyo no jissai, 1988). New York: Productivity Press.

Drucker, P. F., 2010. *The Practice of Management*. New York: HarperCollins.

Ishikawa, K., 1980. *QC Circle Koryo: General Principles of the QC Circle*. Tokyo, Japan: QC Circle Headquarters.

Ohno, T., 1988. *Toyota Production System: Beyond Large-Scale Production*. Cambridge, MA: Productivity Press.

Posteucă, A., 2011. Management branding (MB): Performance improvement through contextual managerial behavior development. *International Journal of Productivity and Performance Management*, 60(5), 529–543.

Posteucă, A. and Sakamoto, S., 2017. *Manufacturing Cost Policy Deployment (MCPD) and Methods Design Concept (MDC): The Path to Competitiveness*. New York: Taylor & Francis.

Chapter 5

MCPD Constant and Consistent Application

The constant and consistent application of the manufacturing cost policy deployment (MCPD) system involves the collection, monitoring, and continuous interpretation of the results of the annual manufacturing cost improvement (MCI) and, implicitly, annual MCI targets and means. The types of data and information collected are for (1) market driven for annual MCI target, (2) profit driven for MCI, (3) driven by annual MCI targets deployment, and (4) driven by annual MCI means deployment (see Section 3.1).

Continuous monitoring and improvement of the MCPD system involves the development of data collection, analysis, and evaluation system at the shop floor level to support people's interest in MCI. At the same time, the continuous reporting of the tangible and intangible results of the annual MCI targets and means takes place.

In this context, in this chapter, answers are given to the following questions: How is the MCPD supported on the long term at all levels of the organization? What is the impact of the MCPD system on the continuous transformation of the manufacturing flow for each PFC? The answers to the first question are addressed in Section 5.1 by presenting the way to monitor and continuously improve the MCPD system, and the answers to the second question are detailed in Section 5.2 by presenting the impact of the MCPD system on the continuous improvement of manufacturing system, more precisely on manufacturing lead time and beyond, on work in progress (WIP), and on material stock.

5.1 Monitoring and Continuous Improvement of the MCPD System

In this section, we will refer to the following: monitoring the current state of key performance indicators (KPIs) related to MCI targets and means; how to quickly recognize, investigate, and eliminate deviations from MCI targets; how to communicate clearly and consistently at each hierarchical level and between hierarchical levels; and effective and efficient meetings to verify the status of MCI targets and means to confirm the annual MCI goal.

5.1.1 MCPD System Information Centers and Horizontal and Vertical Communication

The general values of a consistent communication within the MCPD are the following:

- *Mutual respect*: Being open-minded, having deep understanding of the data and information at the level of each process, and accepting the personality of each person, their ideas, and their opinions
- *Confidence in colleagues' abilities and commitment*: Continuous assertion of what each person thinks and doing what was promised/planned
- *Communication* (clear, coherent, and concise): The precise communication to the target group of a clear message based on actual data and facts (observing the 5Gs for MCI: *gemba* [real/actual place], analysis of the deviations from MCI targets where they occur; *gembutsu* [real/actual activity], focus on data and facts on losses and waste data involved in the deviation from annual MCI targets and means; *genjitsu* [actual/measurable facts], analysis of concrete and measurable losses and waste, cost of losses and waste (CLW), and critical cost of losses and waste (CCLW); *genri* [principles], review of the basic theory and of the physical, chemical, economic, and other principles; and *gensoku* [standards and parameters], continuous verification of values for standards and parameters for CLW and CCLW); providing enough time to talk to each person (MCI daily sessions last for 10–15 minutes; each daily meeting addresses five issues of the day before; each issue

is discussed on the basis of concrete data and facts in order to take promptly the decisions needed)
- *Team spirit*: Provides full support for reaching annual MCI targets and means from the management team and all colleagues

All the MCPD system data and information (annual MCI goal, annual MCI targets and means, annual MCI action plan, annual MCI training plan, and annual MCI performance) are displayed on the *MCPD system information centers* on hierarchical levels (see Figure 5.1).

From the perspective of the MCPD system, the first two hierarchical levels have as main concern the establishment and follow-up of annual MCI targets in order to meet the annual MCI goal (defining control and inspection items for each product family cost [PFC] process with which to support the transformation of the manufacturing flow—*checking MCI results*), and the following three levels are mainly concerned about establishing and following up on annual MCI means targets in order to meet the annual MCI targets (defining control and inspection points for each PFC process—*checking of the causes of MCI problems*).

Viewing this data and information, updating daily (per hours/from shift to shift; *MCI production diary*) helps control the business daily from the perspective of MCI targets and means and recognizes the deviations from the annual MCI goal. Any deviation is analyzed and immediate corrective measures are planned (problem solving—MCI means level 1) or kaizen projects are initiated (MCI means level 2). Trends of this information are displayed and analyzed once a week or monthly using run charts.

5.1.2 Continuously Collecting and Recording the Data and Information

The continuous collection of data and information on the MCPD system is necessary to identify and address issues that contribute to noncompliance of the annual MCI targets and means. An unfulfilled issue or MCI target means that somewhere in the process something did not go as expected. Finding the real causes for not meeting MCI targets requires their control (*control items* and *control point*).

Using the MCI control items/points design, the plant manager should be able to have full visibility over at least two levels below it (at least up to level 3; see Figure 5.1).

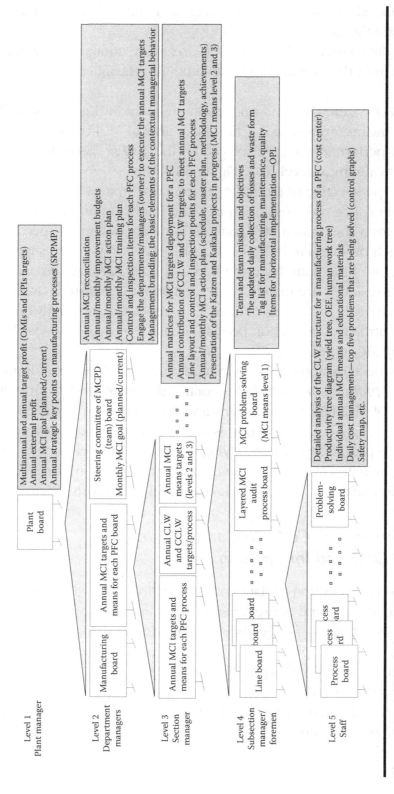

Figure 5.1 MCPD information centers (by levels).

Each management level must have a clearly defined number of *control items and inspection items* both for the fulfillment of the annual external profit expected from sales/production and for achieving the annual MCI goal, as the following:

■ *Plant manager* should check in total between 20 and 50 *control items* for the following:
  – *Annual external profit expected from sales/production*: production numbers, production capacity, price evolution (average), stock level of finished products, units sold daily, unit manufacturing costs, fixed and variable budget expenses, manufacturing overhead costs, manufacturing lead time, production delivery performance (OTIF), labor productivity, labor reduction ratio (moving them to other areas of the factory), value added/person, total number of reduced stock (in terms of money), stock turns, number of claims, number of accidents, innovations in concerned areas, number of innovations (registered, solved), work in progress (WIP) turnover, etc.
  – *Both annual MCI goal and control items*: transformation cost, variable costs (see Section 3.1.2.2); material costs, variable costs (see Section 3.1.2.2); CLW (see Sections 3.1.3.3 and 4.1.1.3) and CCLW (see Section 3.1.3.3); and *MCI control points,* annual % of MCI target touch (% TT) (see Section 4.2.1.2) for annual weaknesses of manufacturing controllable expenses, CLW, and CCLW
■ *Department managers* should check in total between 20 and 50 *control items* for the following:
  – *Annual external profit expected from sales/production*: unit manufacturing costs; fixed and variable budget expenses, from his/her area; manufacturing overhead costs, from his/her area; increase labor productivity, including by reducing the number of people in a certain area/process at the factory (these people being relocated to other factory areas/processes to support consistent company development); scrap cost; overall equipment effectiveness (OEE); output/hour; manufacturing lead times; maintenance costs; number of suggestions; number of claims; number of accidents, innovation, and improvements in the concerned areas; number of innovations and improvements (registered, solved), material yield losses, etc.
  – *Annual MCI goal*: annual MCI targets and means, from his/her area; cumulative effects of MCI means on MCI goal and *control items* (transformation cost, variable costs; material costs, variable costs);

CLW and CCLW; and *MCI control points,* annual % of MCI target touch (% TT) for annual weaknesses of manufacturing controllable expenses, CLW, and CCLW

- *Section manager*: They should have a number of control items ranging between 10 and 15 (annual MCI targets, for the area he/she is responsible for), as the following: number of reduced people from a certain area/process, scrap cost, OEE level, output/hour, mean time between failures (MTBF), number of short stoppages (total duration and frequency), mean time to repair (MTTR), annual MCI means targets, number of improvements (kaizen [registered, solved]), number of suggestions (registered, solved), number of claims, number of process defects and process defect rate, number of accidents, absenteeism rates, *MCI inspection item* (ranging between 10 and 15) (the impact of MCI means targets levels 1 and 2 over the annual MCI targets [or over the annual CLW targets and annual CCLW targets]), *MCI inspection points (% of MCI means target touch [% MTT] over the annual MCI targets)* (ranging between 10 and 15), KPIs for losses (process level), KPIs for waste (process level), KPIs for transformation cost and material costs (process level). Section managers provide the input data for MCI control points at the level of department managers.

- *Subsection manager/foremen*: They should have a number of *control items* ranging between 5 and 12 (annual MCI targets—for the area he/she is responsible for), number of reduced people; scrap cost; OEE level; output/hour; number of machine breakdowns (total duration and frequency); MTBF, number of short stoppages (total duration and frequency); MTTR, annual MCI means targets (improvement in manufacturing lead times, setup times, speed loss, defect rate, yield, labor productivity, transfer time, material handling, etc.); number of improvements (kaizen [registered, solved]); number of suggestions (registered, solved); number of claims; number of process defects and process defect rate; number of accidents; absenteeism rates, etc. Moreover, they collect data on *MCI inspection points (% MTT on the annual MCI targets)* (ranging between 10 and 15); KPIs for losses (process level); KPIs for waste (process level) (Posteucă and Sakamoto, 2017, pp. 119–123); KPIs for transformation cost and material costs (process level).

- *Team leaders*: They should check the following inspection items and points: OEE; output/hour; MTBF; number of short stoppages; MTTR; process/equipment defect rate; MCI means targets (improvements in manufacturing lead times, setup time, speed loss, startup loss, defect rate, yield, etc.); oil/energy consumption; reduction of checking time, time for charging oil and cleaning time, number of suggestions, number of one-point lesson (OPL), etc.

■ *Staff:* They should wish to participate in the annual MCI means targets and be proactive and enthusiastic that they can help solve their own problems or solve the problems of their colleagues. Ideally, annual MCI means targets level 1—problem-solving technique (PST)—should start from the *voices at the floor level* through concrete activities with concrete results, not just through endless discussions. The entire staff needs to be made constantly curious and to live the results in their current state, to go in and see exactly what is in actuality (go and see) and less by the results of the past (statistics and complex calculations).

The *MCPD steering committee* keeps track of all annual MCI check items/points and of all annual MCI inspection items/points and makes monthly, quarterly, semestral, and annual reports to top management on the level of reaching the annual MCI targets and means for each PFC and the annual MCI goal. At the same time, twice a year, it proposes possible adjustments to annual MCI targets and means (in July of the current year and January of the next year).

5.1.3 Management Branding and Managerial Support for the MCPD System

To meet the annual and multiannual MCI means targets, all people must be motivated to do so. In addition to monitoring the tangible effects of the MCPD described in the previous section, it is necessary to obtain a series of intangible effects that support long-term MCPDs, effects that are not quantifiable and cannot be mathematically determined, such as improving the *teamwork* at the level of the MCI implementation teams (setting targets and achieving means), improving *work satisfaction* by presenting the results obtained and by encouraging those who have achieved it, the *complete and timely approach to improvements*, and *increasing confidence* in the work that each person does and in the organization, and the improvement of the need for *knowledge accumulation* and the accumulation of knowledge from all people to cope with MCI issues.

In order to achieve these intangible effects to support the tangible effects of MCPD, there must be a close link with the operators and all the people in the company and beyond, so they will still be happy to meet annual MCI targets by meeting the annual MCI means targets.

Managers need appropriate *management branding* to provide easy, quick, efficient, and less costly systematic improvements (MCI means level 2, kaizen), systemic improvements (MCI means level 3, kaikaku), and problem solving (MCI means level 1, PST). The overall atmosphere created by managers is very important, especially from the perspective of enhancing creativity for the annual MCI means targets.

This atmosphere is created by the continuous development of managers' constant desirable behaviors and by decisions such as

1. The natural rejection of the current situation in the company and explanation of the negative effects of the current state (presenting the need for change as something normal)
2. Ensuring the right people to approach the annual MCI means targets, providing appropriate training programs, and making available all the resources needed for improvement
3. Continuous tracking of annual MCI means for each area of the company
4. Paying attention to the daily MCI management process
5. Trying not to give too much help—developing the capacity for an independent approach to MCI issues by people
6. Setting up groups of three people to address a problem (in case of contradictions of opinions the decision will be taken by majority, 2:1)
7. Assigning people for problems rather than assigning problems to people
8. Addressing rather groups of people who solve MCI problems and not individuals
9. Attracting people from multiple areas of the company to approach a MCI means target, because new ideas cannot be expected continuously from the same people in the area
10. Reducing/eliminating "one man show" behaviors

Therefore, the role of a manager is to meet both production volume requirements and quality assurance and improve the production system, as well as to meet annual MCI targets. The first thing they should do is arrange the manufacturing system so that all anomalies can be noticed immediately, at a glance, with visual control easily understood by anyone and with a visual management integrated into the company's performance management system. Then, the daily behaviors of managers, especially in crisis situations, should not generate confusion, unnecessary comments, irrelevant comments, irrational decisions, unreliable decisions, criticism, and encourage negativity among people in and beyond the company.

5.1.4 Continuous Monitoring of the Annual MCI Goal for Each PFC

Reporting of the current performance of the annual MCI targets and means to meet the annual MCI goal is required for three distinct moments:

- *Planning*: Prepare annual MCI targets and means to fulfill annual MCI goal; prepare annual manufacturing improvement budgets (for existing and new products) and annual manufacturing cash improvement budgets for each PFC; prepare annual action plan for MCI means.
- *Executing*: Engage the workforce to achieve the MCI targets and use information on annual MCI targets and means for decision-making, evaluation of outsourcing opportunities, estimation of intangible effects, analysis and completion of annual MCI training plan, establishment of interdepartmental teams for MCI means (levels 2 and 3; strategic improvements).
- *Reviewing*: Biannual evaluation of the achievement level of the annual MCI targets and means for each PFC and the overall company to make possible any adjustments or to determine the future state of the annual MCI targets and means (in July of the current year and January of the next year)—the annual target profit and annual MCI goal must be achieved even if the annual external profit expected from sales may have negative variations (nonachievement of the annual external profit).

These reports are provided by the MCPD steering committee.

5.2 Impact of the MCPD System on Continuous Improvement of the Manufacturing System for Each PFC

The annual MCI reconciliation of annual MCI targets and annual MCI means (level 3; see Figure 3.4) is used to achieve the annual redesign of the manufacturing flow at the level of each PFC, more precisely the future state of transformation of the manufacturing flow from the perspective of achieving the annual target profit (annual external profit/necessary capacities and, especially, the annual MCI goal). The purpose of this reconciliation is to provide exactly the level needed to meet the *product number, speed and agility, lead time, productivity,* and *product quality ratio.* Fulfilling these goals, at the level of each PFC, provides visible results of improvements, especially for manufacturing lead time, WIP level, and material stock level.

5.2.1 Manufacturing Lead Time

Manufacturing lead time (MLT) represents the total time required to complete an item. As shown in Chapter 3 (see Formulas 3.4 and 3.5), MLT is affected by the number of operations, total cycle time, bottleneck processes/activities, cycle time synchronization at takt time, operation balancing/line balancing, the man*hour level, the WIP level resulting from the setup, and the WIP level resulting from the transfer of stocks and the distances traveled.

Continuous targeting of MLT enhancements by setting and meeting annual MCI targets and means is achieved through the participation of all departments in the company and the external partners involved.

The MLT performance level is given by its synchronization to the factory lead time (*material lead time* [MaLT] and *delivery lead time*).

Often, results are visible for the following:

1. *The number of operations*: Reduction/optimization (man*hour is optimized/reduced to minimum), correct sizing of the required labor hours; the right dimensioning of direct and indirect labor costs.
2. *Bottleneck processes/activities*: Their identification and monitoring for each manufacturing area; fitting them into the takt time needed for manufacturing; continuous reduction of bottleneck-related CLW and CCLW.
3. *Cycle time synchronization at a takt time*: Cycle time for processes and equipment is optimal to reach the manufacturing takt time; improving ergonomics; the operations and processes of new products and new equipment fall into the manufacturing takt time; continuous reduction of CLW and CCLW due to nonsynchronization to customer demand takt time.
4. *Operations balancing/line balancing*: Operation, process, and lines are balanced with each other to ensure the level of takt time; manufacturing takt time is synchronized to supply chain takt time; manufacturing takt time is synchronized to the customer demand takt time, the continuous reduction of CLW and CCLW related to unbalanced operation, process, and lines.

5. *WIP level resulting from setup*: Setup time is minimized (sometimes under 10 minutes), minimal/maximum planning takes into account the setup time; the cost of setup time losses is continuously reduced, as well as the CCLW-related setup.

6. *WIP level resulting from the transfer of stocks*: Optimization of transfer speeds, automation of the transfer between processes; reduction of manual transfer (elevators, rotating tables, etc.); reducing/eliminating heavy stock transfers; developing the transfer ordering system; transfers take account of the production mix; eliminating safety risks for reducing transfer time; continuous reduction of CLW and CCLW associated with WIP from the transfer.

7. *The distances traveled*: Reducing distances on the layout; reducing the length of transfer devices; reducing the length of lines; continuous reduction of CLW and CCLW associated with distances traveled.

8. *OEE/OLE (overall line effectiveness) for equipment*: Continuously improved to fall within the current admissible losses and for the continuous reduction of CLW and CCLW associated to OEE/OLE.

Therefore, by continually improving the MLT, productivity and flexibility are increased, the availability of more orders and higher volumes increases, the manufacturing costs decrease (both material costs and transformation costs; CLW/product), and the cash flow is improved.

5.2.2 Work in Progress

WIP, also called work in process, is an inventory that has entered the manufacturing process, is no longer in the stock of raw materials, is not yet finished, and has reached the finished product category. As presented in Chapter 3 (see Formulas 3.4 and 3.5), WIP is a component part of MLT. Sometimes the time spent by an item in the WIP state may be 95%–98% of the MLT. The lower the WIP, the shorter the MLT, and the productivity increases.

The visible results obtained with the MCPD system from the WIP perspective, obtained through annual MCI targets and means, are the following: minimizing the WIP value; the cost level of WIP is known on a continuous basis; CCLW levels caused by WIP are continually reduced; the minimum and maximum levels of WIP are observed and take into account the setup time, transfer time, production changes, and production constraints; WIP minimum is continually reducing and takes into account the quality constraints that generate WIP; continuous implementation of automation to reduce transfer costs (MCI mean level 3, kaikaku); redesigning workplaces to achieve WIP reduction (MCI mean level 3, kaikaku), paralleling parallel line operations are eliminated; the distance between two operations is minimal; non-value-added activities are continually reduced and/or eliminated; and between two operations there is only one item.

Therefore, amid the continuous reduction of WIP through the MCPD system, productivity increases, MLT reduces, and the manufacturing costs decrease (CLW/product).

5.2.3 Material Stock

Material stock refers to stockpiles of raw materials, auxiliaries, parts, and components required for the manufacturing processes and inventory-related equipment, lines, or processes that are not WIP. The use of any stock material is to turn as quickly as possible into finished products and then to be sold.

The visible results obtained with the MCPD system in terms of stock material, achieved through annual MCI targets and means, are as follows: The minimum and maximum stock is stable and the minimum stock is continually reduced; the material costs are reduced together with CLW and CCLW due to a lack of material; noncompliant products are continuously reduced; all materials are identified and accessible; the inventory of equipment, lines, or processes is continuously known and continuously removes landmarks that are no longer in line with the current production plan; the capacity of the equipment, the lines, and the processes is synchronized with the material refueling system (times synchronization— material handling activities).

Therefore, due to the continuous reduction of the stock material through the MCPD system, the material costs and CLW/CCLW are reduced, productivity is increased, MLT is reduced, and, finally, the material stock days decrease.

Reference

Posteucă, A. and Sakamoto, S., 2017. *Manufacturing Cost Policy Deployment (MCPD) and Methods Design Concept (MDC): The Path to Competitiveness.* New York: Taylor & Francis.

MCPD PRACTICAL IMPLEMENTATIONS AND CASE STUDIES

Chapter 6

Applications of the MCPD System

In this last chapter, two applications of the MCPD system in two real manufacturing companies from two different industries, namely, the *manufacturing and assembly industry and processing industry*, are presented. For both applications, the following are addressed: the main actions, the main activities, and the main challenges for starting the MCPD implementation. At the same time, the MCPD steps and two real case studies for each company are presented. (MCI means only two out of about 12 annual and multiannual strategic productivity improvement projects are held to meet the annual MCI goal and multiannual internal profit and target profit.)

For the first company, manufacturing and assembly industry ("AA Plant"), in Section 6.1, MCI by improving the equipment setup and adjustment time and associated costs (with KAIZENSHIRO) (case study 1) and MCI by increasing productivity with the replacement of bottleneck equipment (with KAIKAKU) (case study 2) are presented. For the second company, process industry ("BB-Plant"), in Section 6.2, MCI by improving operators skills to sustain quality (with KAIZEN) (case study 1) and daily MCI management—cost problem solving for lubricants consumption for one of the equipment (with A3; as an activity associated with the multiannual strategy to reduce the lubricant consumption of the equipment)— (case study 2) are presented.

The results of applying the MCPD system to the two companies, by continually targeting productivity improvements through the need to meet annual MCI goal and multiannual internal profit and profit targets, have been fully achieved. The planned and achieved average annual manufacturing unit cost reductions were about 6%.

Therefore, *the MCPD system starts and ends with the multiannual target profit and associated multiannual target productivity needed to be met.*

6.1 "AA-Plant": Manufacturing and Assembly Industry

The following is a case for implementing a pilot project for the MCPD system at AA-Plant. It will emphasize the presentation of the main core activities, basic steps, results, and challenges.

AA-Plant is a multinational company operating in the automotive industry. AA-Plant is one of the top five companies in the group, with approximately 900 employees. Following the initial analyses, three *product family costs* (*PFC*) were identified (PFC1, PFC2, and PFC3—225 employees).

The main objective of the plant manager at AA-Plant was to counteract the current deficiency of unit cost reduction for the next three years for C1, maintain the planned annual profit level, and ensure a smooth manufacturing flow transformation for PFC3 through increasing productivity and synchronizing to C1's needs and the market in general. To meet the two objectives of the MCPD project (cost reduction and maintaining profit), the top management at AA-Plant was aware that PFC3 required a radical change in the current way of reducing manufacturing lead time (MLT), increasing synchronization processes at takt time (especially between the equipment and the PFC3 assembly line), reducing work-in-process (WIP) stocks, creating One-Piece Flow in as many areas on the flow, if not on the entire manufacturing flow, maximum reduction of the material stocks, and, especially, the reduction of cost of losses and waste (CLW) and critical cost of losses and waste (CCLW).

For this plant, the AA-Plant manager decided to set up a team to implement the *pilot project of the MCPD system.* Champion MCPD was plant manager at AA-Plant. The *Manufacturing System Transformation by MCPD* team was made up of production manager, maintenance manager, production engineering manager (including some cost-related tasks), quality assurance manager, production control manager, managerial and cost accounting manager, HR manager, development and design engineer (including special tasks on the new products profitability), production line team leader, two experienced/polyvalent operators for major operations, improvement specialist (also in charge of the MCPD project management), and a consultant in the MCPD system. For each member of the MCPD team, tasks were assigned to the project, and each member formally selected a stable working team consisting of three to five people from different departments, including members of the MCPD team. The team participated in the MCPD system training and workshop program (Basic MCPD—four days).

6.1.1 Establishment of MCPD System: Actions, Activities, and Challenges

The pilot project for the application of the MCPD system for PFC3 lasted 15 months (beginning in October of year "N − 1"). Following the results of the pilot project, it was subsequently decided to adopt the MCPD system throughout the company.

In the last three months of "N − 1," organizational activities leading to the start of the MCPD system at AA-Plant for PFC3 have been carried out, such as PFC definition, product structure, capture price evolution, price benchmark analysis (identification of current and future pressure of prices from customers and competitors), current and future competitive gap, long-term sales, set targets selling price, long-term and annual realistic manufacturing profit, internal and external profit, annual manufacturing cost improvement (MCI) goal (for the year "N"), cost system analysis and the needs for MCPD, understanding the manufacturing flow from the perspective of the MCPD system (losses and waste analysis, identifying how to determine the CLW and CCLW), scenarios for gaining profit directly from internal processes, and annual reconciliation to set MCI targets.

In the first 6 months of year "N," activities for MCPD planning for PFC3 have been carried out, such as MCPD transformation draft plan, internal and external communication of the MCPD system purpose, setting annual MCI targets and means (steps 1 and 2 of MCPD), developing annual manufacturing improvement budget (AMIB) for the existing products and for a new product, developing annual manufacturing cash improvement budget (AMCIB) (step 3 of MCPD), and developing an annual MCI action plan (step 4 of MCPD).

Starting with the last six months of year "N," running the activities and actions of the annual MCI means (step 5 of MCPD) was started— the assessment of MCI performance (step 6 of MCPD) and daily MCI management (step 7 of MCPD) and monitoring and continuous improvement of the MCPD system.

6.1.2 MCPD Implementation Steps: From Productivity Strategies to Annual MCI Targets and Means

Here are the main activities and results of applying the MCPD system to AA-Plant in PFC3.

6.1.2.1 Step 1: Context and Purpose of MCI

As a result of the activities carried out in the last three months of the year "N − 1," the following information resulted.

6.1.2.1.1 Context and Purpose (Background)

In PFC3 (dedicated manufacturing flow), four product types were manufactured for eight key customers. The planned three-year sales trend ("N − 3," "N − 2," and "N − 1") for PFC3 was slightly rising. This trend will be retained for the next three years if the signed contracts are renewed. Like any automotive company, AA-Plant has a culture of improvement at all levels, and AA-Plant production management was focused on Lean Six Sigma methodologies. However, top management felt the need to increase the level of competitiveness by cost and profit, and to know their scientifically determined strategic cost reserve on a continuous basis through projects to improve productivity and quality. Currently, top management at AA-Plant has set a 5% annual cost reduction as an objective. Moreover, the regular audits of some key clients in PFC3 started to focus on the need to continuously reduce unit costs, especially as long-term agreements with them were about to expire. The requirements of the eight key clients of PFC3 were increasingly difficult to meet, especially in terms of costs. But, however, top managers at AA-Plant were willing to keep their business with all eight customers and especially with the basic C1 customer. C1's requirements for PF3 for the next three years ("N + 1," "N + 2," and "N + 3") were representative of other customers and aimed at reducing unit costs by 5.5% per annum (with the possibility of reaching a 6% reduction starting with year "N + 2"), increasing production delivery performance (OTIF, On-Time In-Full) from 96% to 99% in the year "N + 3," the reduction of FLT from 28.75 days to less than 10 days in year "N + 3," and reduction of MLT from 5 to 2.5 hours. The plant manager at AA-Plant was determined to continue business with C1 since C1 accounted for 37% of PF3's turnover, had innovative products, implicitly, at the beginning of the life cycle and is forward-looking, and offered important orders to other product families (PFC1 and PFC2).

Based on these current business conditions, including requirements from the C1 representative customer, AA-Plant's top management set the following corporate goals for the following 5 years: (1) increase global market share, (2) decrease the delivery time, (3) decrease the manufacturing cost, (4) increase the manufacturing profit, (5) improve quality, and (6) the continuous development and implementation of new profitable products. Starting from the need to reduce costs and increase profits, the topic of the MCPD pilot project for year "N" was to achieve the following two main objectives for the next three years for PFC3:

1. Reducing unit manufacturing cost by at least 5% per year
2. Providing annual target profit for PFC3 of at least 10% per annum (both from annual external profit expected from sales/production, especially from the annual MCI goal)

For this, all systematic improvement activities and systemic improvement actions will be aligned to these two key strategic objectives.

Prior to adopting the MCPD system, AA-Plant had a culture of improvements based on improving quality, productivity, and then costs (ultimately, as a result of all improvement projects). Therefore, before the implementation of the MCPD system at AA-Plant, top management was guided by the relevance of the quality and timing of processes against cost.

The turnover expected for the next three years for C1 was increasing, as is the case for the other PF3 customers; as we have already said, the need for flexibility tended to increase (the number of changeover required by the orders received from PFC3 customers was more and more frequent) as was the number of new products launched in production; and, lately, some equipment replacements (systemic improvements—kaikaku) did not have the expected results from the perspective of unit cost reductions. Consequently, amid the overcapacity in growth (about 28%), some of the cost centers and, implicitly, the unit manufacturing cost had a tendency to increase direct labor, electricity, handling material, raw materials, consumables, spare parts, etc.

C1's requirement to reduce costs by 5% over the year "N + 1" (with the possibility of increasing to 6% from the year "N + 2") was considered the biggest challenge as it was considered to have already made great efforts in the past to reduce costs and that there are not so many opportunities in current manufacturing processes. The plant manager's expectations for the annual cost reductions were 4.5% for the next three years. The negative impact on the profit margin of the 1% difference (the difference between

5.5% and 4.5% in the year "N + 1") and 1.5% (the difference between 6% and 4.5% in the year "N + 2" and the year "N + 3") in the profit for the next three years was estimated at a total of approximately $1,200,000 (considering the trend of potential sales volumes). He tried to negotiate this percentage of cost reduction, but the answer from the client C1 was as follows: "After the past two years of doing business together, it's time to share your profits, because AA-Plant has gained experience in manufacturing and delivering our products, even if they are sometimes innovative."

6.1.2.1.2 Manufacturing Cost Key Points and Manufacturing Key Points

The designated team analyzed the main key performance indicators (KPIs) for PFC3 in the past ("N − 3," "N − 2," and "N − 1") and the current year targets ("N") to identify the main phenomena and principles of opportunity manifestation for MCI (see Table 6.1).

As noticed, the analysis aimed at identifying the trends of the main historic KPIs and then making their projections for the next three years. The purpose of analysis of the four coordinating (driven) directions for meeting the two major objectives of the MCPD project (transformation triggers for annual MCI goal) was to help establish the annual MCI targets and means for the next three years and especially for the "N" year starting from the current "N"-year targets (established before the MCPD system approach). The year "N–2" was the reference year for manufacturing costs, a year in which accentuated concerns for cost reduction have begun.

6.1.2.1.3 Identifying the Structure and Evolution of Average Unit Prices

This was achieved in connection with the structure and evolution of the unit manufacturing costs and unit profit. The *following scenarios for MCI targets* were set for year "N":

- Reduction of the annual average price from $9.87 to $9.17—more exactly by $0.7 (or 7%, based on customer C1's requirements to reduce unit average costs for year "N" by 5.5%)
- Planned profit growth from $0.94 to $1—more exactly by $0.06 (or 6.38%; taking into account the need for additional profit, more than 9.52% of the price at the beginning of year "N," to recover part of the nonperforming profits in the year "N − 1")
- Reducing transformation costs from $2.10 to $1.92 (or by 8.57%) and material costs from $6.83 to $6.24 (or 8.63%); by exploiting opportunities to reduce manufacturing costs from current processes by increasing productivity

Table 6.1 Evolution of KPIs for the PFC3

MCPD	OMIs	KPI	Unit	N − 3	N − 2	N − 1	N
Market driven for MCI	1.	Sales turnover (for PFC3)	$	16,547,517	16,987,783	18,412,947	21,650,804
		Production numbers	Products	1,445,890	1,511,784	1,735,465	2,181,695
		Stock level of finished products	Products	12,041	11,784	12,032	12,363
		Price evolution (average)	$	11.35	11.15	10.54	9.87
	2.	Market Share for target area (for PF3)	%	47.5	48	50	52
Profit driven for MCI	3.	Manufacturing profit (for PFC3)	$	1,151,765	1,752,103	1,899,093	2,057,316
		Unit profit	$	0.79	1.15	1.09	0.94
		Production capacity	Number	1,767,987	1,835,400	2,392,017	2,590,785
		Scrap ratio	% ($ base)	2.5	2.2	1.55	1.5
		Rework for assembly	%	14	9	7	6
Driven by MCI targets deployment	4.	Manufacturing cost (for PFC3)	Index (% based)	N/A	1	94.50	89.30
		Total manufacturing cost	$	15,395,751	15,235,680	16,513,853	19,593,488
		Manufacturing unit cost	$	10.56	10.00	9.45	8.93
		Transformation costs	$	3.06	2.80	2.55	2.10
		Material costs (raw materials and auxiliary materials)	$	7.50	7.20	6.90	6.83

(Continued)

Table 6.1 (*Continued*) Evolution of KPIs for the PFC3

MCPD	OMIs	KPI	Unit	N – 3	N – 2	N – 1	N
Driven by MCI means deployment	5.	**Total plant lead time (for PFC3)**	Days	33.7	31.27	29.75	26.23
		Manufacturing lead time	Hours	7.8	6.5	6	5.5
		Total cycle time	Hours	0.63	0.58	0.55	0.50
		Production delivery performance (OTIF)	%	92	95	94	95
		Major accident	Number	0	2	2	0

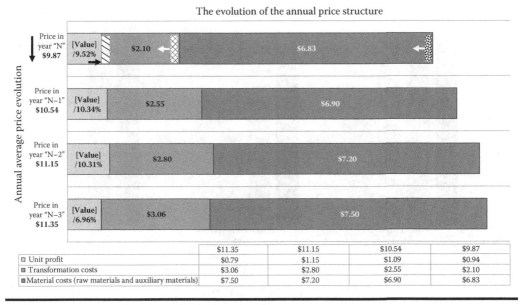

Figure 6.1 Structure and evolution of prices in connection with the structure and evolution of unit manufacturing costs and unit manufacturing profit for the PFC3.

Thus, the unit manufacturing cost reduction target for the year "N" was $0.76 (or 8.51%, the difference between $8.93 and current cost of $8.17) not only to meet both the cost reduction requirement imposed by the customer C1 of 5.5% and the need to increase the unit profit, but also to mitigate some of the risks on the evolution of costs that may occur in year "N." This represents *top-down approach for annual MCI targets* (Figure 6.1).

Since the target prices, target profit, and type and quality of raw materials and auxiliary materials (including their price) are largely imposed by entities outside AA-Plant, the MCPD team analyzed the structure and evolution for *transformation costs* to look for directions of fulfilling the two project objectives (Figure 6.2).

Furthermore, within the *transformation cost* structure, the following structure of the *Manufacturing Overhead* ($0.86 for year "N − 3"; $0.73 for year "N − 2"; $0.66 for year "N − 1"; and $0.51 for year "N") was identified: maintenance costs/spare parts, utility costs, and other overhead costs (Figure 6.3).

As can be seen in the *Manufacturing Overhead* structure, a significant part is represented by other overhead costs.

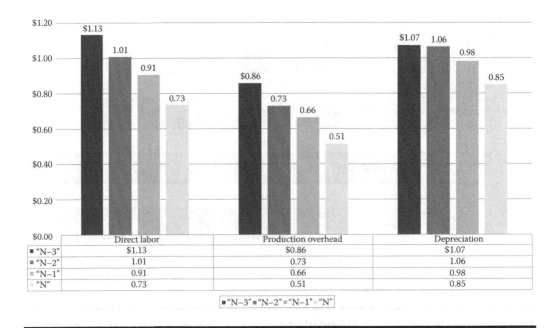

	Direct labor	Production overhead	Depreciation
■ "N–3"	$1.13	$0.86	$1.07
■ "N–2"	1.01	0.73	1.06
■ "N–1"	0.91	0.66	0.98
"N"	0.73	0.51	0.85

■ "N–3" ■ "N–2" ■ "N–1" ■ "N"

Figure 6.2 Structure and evolution of transformation costs for the product family of the PFC3.

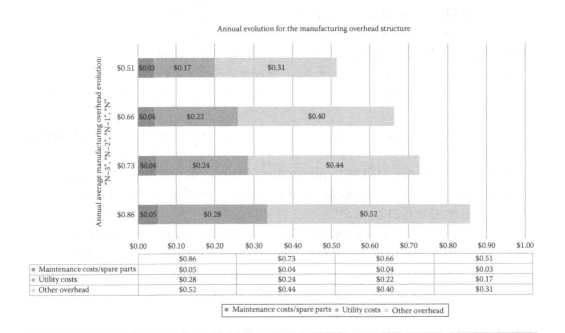

	$0.86	$0.73	$0.66	$0.51
■ Maintenance costs/spare parts	$0.05	$0.04	$0.04	$0.03
■ Utility costs	$0.28	$0.24	$0.22	$0.17
Other overhead	$0.52	$0.44	$0.40	$0.31

■ Maintenance costs/spare parts ■ Utility costs ■ Other overhead

Figure 6.3 Structure and evolution of manufacturing overhead for the PFC3.

The basic questions of the MCPD team were as follows:

1. What is the current CLW level taking into account the bottleneck cycle time implications over the entire manufacturing flow of PFC3 and the WIP level of the setup and the transfer times between processes for all areas of production?
2. What are the costs of non-one-piece flow and noncontinuous flow and where are they located to interfere with them (CCLW identification)?
3. How to tackle direct labor cost reduction by redesigning current workplace design without affecting the quality of products (methods design concept [MDC])?

Based on these first cost structure analyses, the MCPD team has begun to realize that within each cost element, both *material costs* (raw materials and auxiliary materials), *transformation costs, and especially variable manufacturing overhead*, there are many opportunities for cost reductions. From the perspective of the MCPD system, the determination of opportunities for MCI, based on *manufacturing cost key points*, has been made aware at the direct labor level and especially at the manufacturing overhead level. However, they still could not locate and measure exactly all the opportunities to improve the unit manufacturing costs in all PFC3 processes. For this, the team analyzed the manufacturing flow to understand the flow from the perspective of the MCPD system for both the *manufacturing and assembling areas (assembly line area)*. In fact, starting from manufacturing cost key points, the MCPD team sought *manufacturing key points* to determine the real opportunities for MCI by improving productivity.

For the manufacturing area, we analyzed the *technological flow for injected and painted parts (processes in area 1 and processes in area 2)* to understand the current state of MCI opportunities and, implicitly, the current system of computing and collecting manufacturing costs at the centers' cost level. Further, for example, for the technological flow of injected parts (as a cost center), the following technological operations were analyzed: (1) granulation drying, (2) injection molding preparation process, (3) plastic injection process, (4) unloading of injected parts, (5) packaging, (6) storage, and (7) transfer to the technological flux of dyeing injected parts. For each of the seven technological operations, the phases of the operations and their implementation (manual or with equipment/machines) were analyzed to identify the types of times in which the costs and the costing of each cost center are lost (generally, each equipment is a cost center). For example, for the (1) granule drying, the following phases were analyzed: material feed, unsealing bags with material, bulk feeding with material, drying temperature programming, automatic inspection. For each phase, the main types of time lost by people and equipment and the types and quantities of unnecessarily consumed materials and utilities were identified in order to be able to quantify subsequently the cost for each cost center. *The method and frequency of collecting losses and waste for each PFC3 process have been determined.* The MCPD team was aware that the *technological flow for painted parts (processes in area 2)* was a *manufacturing cost key point* as it represented *bottleneck areas* for PFC3, but the impact of CCLW and CLW on all PFC3 could not be known exactly. Similarly, the MCPD team knew the *bottleneck processes from each area (manufacturing cost key points)* but still could not scientifically determine their impact on the opportunities to meet the two MCPD project targets (the summed costs generated by the bottleneck processes—CCLW).

6.1.2.2 Step 2: Costs Strategy into Action: MCPD Alignment for Setting Annual MCI Targets and Means

Starting from the MCI profit scenarios described earlier and the 8.51% cost reduction target (*top-down approach for annual MCI targets*), the following measurements during the first six months of year "N," the *actual losses and waste (bottom-up approach)*, were identified (Posteucă and Sakamoto, 2017, pp. 116–132):

- Lost times in people's work
- Lost times in operating the equipment
- Unnecessary use of materials and utilities for PFC3

Such measurements were made on all technological flow for injected and painted parts and for the assembly line area for PFC3.

Based on the standard costing system, the team could turn every minute and waste into costs for each cost center. At the same time, an ideal cost could be set if all unnecessary costs behind unnecessary times and behind useless material consumption (*zero cost of losses and waste*) were to be eliminated from all PFC3 processes and a MCI level possible to be reduced in the next period could be set (*annual MCI goal*) by meeting the annual MCI targets at process levels and by systematic improvements (*annual MCI means level 2—kaizen*) and systemic improvements (annual MCI means level 3—kaikaku) to achieve the target unit cost reduction rate of 8.51% for year "N" and the annual MCI goal. Part of these improvements relates to MCI, another part aims to reduce the cost of purchasing of raw materials and materials created outside the manufacturing (in the supply chain, considering that there are no other costs such as distribution costs, environmental costs, costs after the end of product life cycle, marketing costs, etc.), and another part aims at designing and/or redesigning products. The focus area of MCPD is to meet the *annual MCI targets to achieve the annual MCI goal.*

As a result of the transformation of losses (for equipment, for human work, and for materials/energy) and waste (stocks) in costs for each process/cost center in each of the three production areas (processes in area 1, processes in area 2, and assembly line area), for the first six months of year "N" for PFC3, total unnecessary costs of $3,085,974 (consisting of equipment $2,016,684, human work $495,916, materials/energy $337,297, and waste/stocks $236,077) were highlighted. Taking into account the specificity of the activities for the next 12 months, the relatively constant level of planned production volumes, and the evolution of internal and external costs (from suppliers), the MCPD team has set a *CLW* level for the next three years of approximately $6,200,000 (total internal profit for three years). For the next 12 months (six months until the end of year "N" and the first six months of year "N + 1"), a total MCI target of $514,616—unnecessary costs to eliminate—(consisting of equipment $341,492, human work $26,231, materials/energy $131,154, and waste/stocks $15,739) was established. For the next six months, by the end of year N, a MCI target of $262,308 (consisting of equipment $175,746, human work $13,116, materials/energy $65,577, and waste/stocks $7,869) was set. At the same time, the team determined that in business conditions known at that time, the *total internal profit* for the next three years is about $6,200,000 and that the acceptable annual level of the MCI goal is about $500,000 (or *annual MCI goal*). This is considered as the multiannual tangible stake of MCPD (*gaining hidden reserves of profitability by productivity*) and the main command for productivity need for PFC3 (see Table 6.2).

Figure 6.4 shows an example of a *CLW determination* for the *technological flow of injected parts* (processes in area 1, process "e1"). As can be seen, for the first six months of the "N" year of measuring losses and waste for process "e1" and transforming them into costs, resulted a CLW of $22,326.6, which is the maximum amount for MCI at that time. That amount is set for annual MCI targets in correlation with all other processes along the

Table 6.2   Forecasts for Annual MCI Goal and Setting Annual MCI Targets for Each Area of PFC3

				N	N + 1	N + 2	N + 3
Annual MCI Goal—Predicted			$	498,588	494,095	524,616	499,592
Annual MCI targets on each area	Equipment CLW	Processes in area 1	$	50,108	49,656	52,724	50,209
		Processes in area 2	$	250,540	248,282	263,619	251,045
		Assembly line area	$	33,405	33,104	35,149	33,473
	Human work CLW	Processes in area 1	$	2,493	2,470	2,623	2,498
		Processes in area 2	$	6,232	6,176	6,558	6,245
		Assembly line area	$	16,204	16,058	17,050	16,237
	Materials/ energy CLW	Processes in area 1	$	105,950	104,995	111,481	106,163
		Processes in area 2	$	12,465	12,352	13,115	12,490
		Assembly line area	$	6,232	6,176	6,558	6,245
	Waste (stocks) CLW	Processes in area 1	$	748	741	787	749
		Processes in area 2	$	748	741	787	749
		Assembly line area	$	13,462	13,341	14,165	13,489

manufacturing flow so that the *annual MCI goal* is met (of 8.51%, *top-down approach for annual MCI targets*).

To meet the annual MCI targets for process "e1," in correlation with all other processes of PFC3, targets are set for losses and waste, respectively, for the following: 107 hours of equipment losses, 53 hours of losses for human work, $739 cost of losses related to material and energy, the 3,100 pieces represent waste/stocks (including for WIP from the 30-piece set and WIP from move for 70 pieces, the parts being determined as a lost production opportunity for WIP times). Continuous measuring of losses and waste for all processes in PFC3 underlies the establishment of annual MCI means feasible for the fulfillment of the annual MCI targets (see Figure 6.5).

Moreover, through these continuous measurements of productivity for each process and process activities, loss and waste described in Figure 6.4, *bottleneck processes*, and, implicitly, *bottleneck* activities that spread losses and waste along the entire manufacturing flow can be determined, making the associated bottleneck costs increase. As can be seen in Figure 6.5, starting from the current MCI maximum target of $22,326.6,

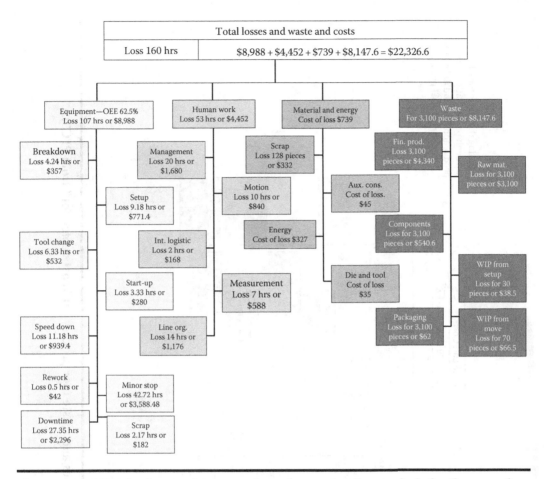

Figure 6.4   CLW: background on manufacturing cost policy analysis for the manufacturing process (per process "e1"/cost center "e1" from PFC3).

the targets for losses and waste are set to achieve the *CLW* target by improving process productivity (especially for bottleneck activities). For example, knowing the setup loss of 9.18 hours or $771.4 and WIP from setup loss for 30 pieces or $38.5, it can be determined that the setup activity is a *bottleneck* activity that causes costs at least for WIP from setup loss in "process e" but also along the entire manufacturing flow. Therefore, there is no point in planning activities to improve setup-related WIP (it is an effect), but rather setup improvement activities (is a cause). Therefore, the impact of *setup activities in CLW* is much higher along the entire manufacturing flow for PF3 than only in the "process e," if all nonproductive times and all the stocks resulting along PFC3 are taken into account. These are CCLW. Unplanned setup activities and setup time exceedance are the two causes of WIP from setup loss over

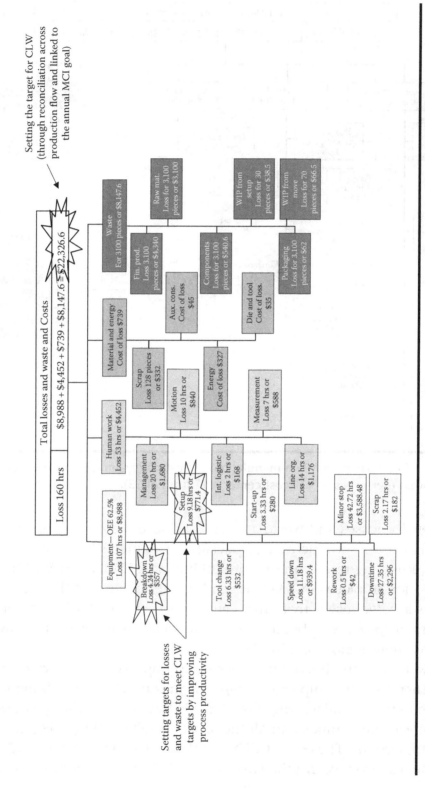

Figure 6.5 Setting the targets for CLW and for losses and waste (per process "e1"/cost center "e1" from PFC3).

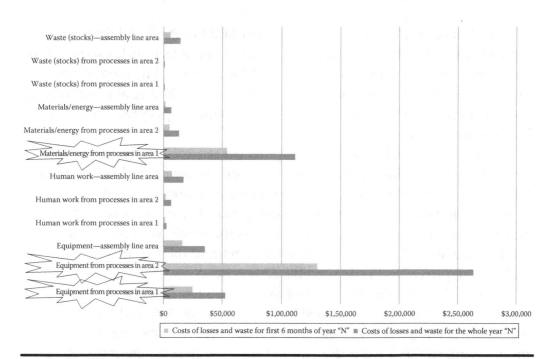

Figure 6.6   Annual MCI targets on each area based on CLW and CCLW—from all PFC3 processes.

the current acceptable WIP (*Waste/Stocks Elasticity on Losses*). Reducing this overrun of standard WIP is what must be seen in concrete terms after the kaizen projects for setup.

Going forward, with the CLW structure and *bottleneck processes* reasoning that dissipates losses and waste (and related costs) across the entire manufacturing flow for PFC3 for the first six months of year "N" (CCLW), the projections for CLW for the whole year "N" for all areas of PF3 are established.

Figure 6.6 shows how to set the annual MCI targets on each area based on CLW and CCLW from all PFC3 processes. As can be seen, efforts to improve MCI for the next six months of year "N" will be focused especially on the following: (1) equipment from processes in area 1, (2) equipment from processes in area 2, and (3) materials/energy from processes in area 1.

At the same time, this stratification process has highlighted that 54% of the CLW for the whole year "N" is in processes in area 2. Thus, 54% of *uncovering hidden reserves of MCI and, implicitly, profitability* is located in processes in zone 2 (*technological flow for painted parts*). Thus, efforts to improve productivity will be directed to this end. However, the MCPD team will develop plans to meet the annual MCI targets for all three areas (technological flow for injected parts, processes in area 1; technological flow for painted parts, processes in area 2; and for assembly line area, processes in area 3), so that a participation share in the annual MCI goal is set for each area by meeting annual MCI targets—by involving all people at all hierarchy levels of the AA-Plant.

Based on the number of products processed in each of the three areas (technological flow for injected parts, processes in area 1; technological flow for painted parts, processes in area 2; and for assembly line area, processes in area 3), and each cost center (each equipment being a cost center), the percentage of total unit manufacturing costs that could be reduced by meeting the annual MCI targets in order to achieve the two objectives of the MCPD pilot project in PFC3 could be determined.

Figure 6.7 shows the structure of average unit price evolution for years "N − 3" ($11.35), "N − 2" ($11.15), "N − 1" ($10.54), and "N" ($9.87). As can be seen in year "N," the total annual opportunities for MCI for a product unit of $0.76 were determined (*annual improvement target for unit costs of waste/stocks of $0.31/unit* and annual improvement target for unit costs of losses for transformation cost of $0.45/unit) of the total unit price of $9.87. Total unit costs of waste/stocks for future improvement of $1.65/unit and total unit costs of losses for transformation costs for future improvement of $0.18/unit represent the current total reserve to reduce unit cost ($1.83/unit) that will be addressed in the coming years, also taking into account the current conditions of those years (precisely the actual cost and CLW/CCLW level).

Further on, Figure 6.8 shows the percentage structure of average unit price evolution for "N − 3," "N − 2," "N − 1," and "N." As can be seen, for the year "N," 7.7% of the price structure is associated with an *annual MCI goal* (*annual improvement target for unit costs of waste/stocks* of 3.1%/unit and *annual improvement target for unit costs of losses for transformation cost* 4.6%/unit). In the next two to three years, this percentage will tend to increase as the MCPD team experience has increased in identifying *CLW* and *CCLW* at each process level. It can be considered that approximately 30%–35% of the cost structure can be associated with *opportunities for MCI* for the first two to three years of application of the MCPD system. Then, this 30% percentage tends to fall amid the implementation of improvement solutions (annual MCI means—levels 2 and 3).

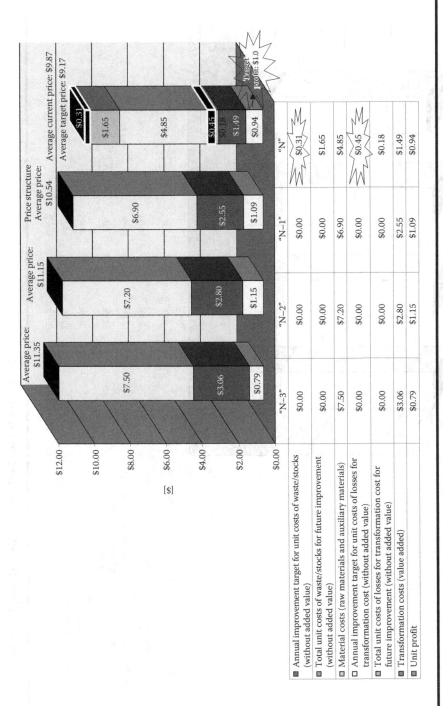

Figure 6.7 Opportunities for MCI at product unit level (in $).

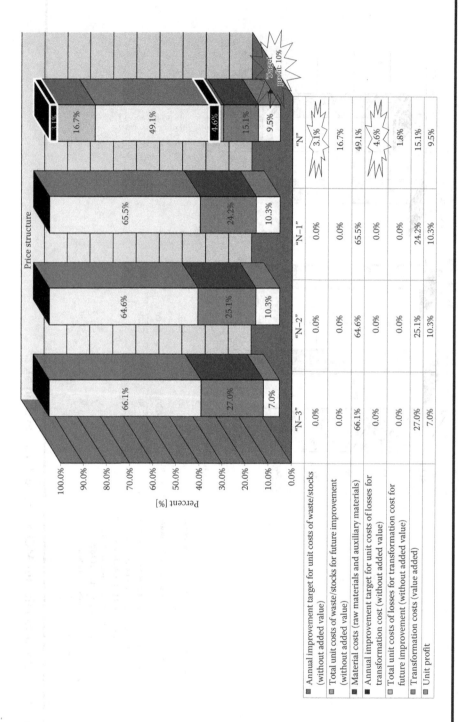

Figure 6.8 Opportunities for MCI at product unit level (in %).

	"N−3"	"N−2"	"N−1"	"N"
■ Annual improvement target for unit costs of waste/stocks (without added value)	0.0%	0.0%	0.0%	3.1%
▢ Total unit costs of waste/stocks for future improvement (without added value)	0.0%	0.0%	0.0%	16.7%
■ Material costs (raw materials and auxiliary materials)	66.1%	64.6%	65.5%	49.1%
■ Annual improvement target for unit costs of losses for transformation cost (without added value)	0.0%	0.0%	0.0%	4.6%
▢ Total unit costs of losses for transformation cost for future improvement (without added value)	0.0%	0.0%	0.0%	1.8%
▨ Transformation costs (value added)	27.0%	25.1%	24.2%	15.1%
▨ Unit profit	7.0%	10.3%	10.3%	9.5%

6.1.2.2.1 Alignment for Setting Annual MCI Means

By identifying the difference between manufacturing costs and annual MCI targets that could be achieved in the next period (the last six months of year "N" and the first six months of year "N + 1"), the MCPD project team developed different scenarios to address current capacity increases, in particular, equipment and people, through systematic process improvements (annual MCI means level 2—kaizen projects) and some projections on some process systemic improvements (annual MCI means level 3—kaikaku projects). These scenarios were aimed at establishing annual MCI means targets to reconcile the pressure on annual MCI targets coming from top to bottom (or top-down approach of $0.76/unit or 7.7%/unit) with total current availability to achieve the annual MCI targets (or bottom-up approach of $2.59 or 26.24% with $0.63 or 6.4% availability for MCI). Therefore, starting from the need for annual MCI goal, targets for losses and waste have been set along the entire manufacturing flow of PFC3. The setting of annual MCI means targets along the entire manufacturing flow for PF3 has primarily taken into account the bottleneck processes that spread losses and waste both upstream and downstream of the entire manufacturing flow (CCLW). On the basis of systematic improvements (kaizen) and rethinking of working methods (MDC), the investment on capacity increase requested by the customers was reduced on average by 14% for year "N" and with projections of at least 20% for the next three years. Following the reconciliation of the annual MCI targets and means (through MCICP—negotiating MCI targets and means at all AA-Plant levels to meet the MCI goal), they were communicated at each level in AA-Plant for each employee to prepare the actions and the activities required to reach the annual MCI targets for each area in conjunction with the other areas so that each employee contributes to the two objectives of the MCPD system. At the same time, the annual MCI targets and means summary was presented to the C1 customer for possible adjustments. Customer C1 did not request any additional adjustment. Therefore, the primary concern of the MCPD system is to meet the annual MCI targets and means at the process level to meet the annual MCI goal (including annual target profit) and to meet the target price imposed by customers.

6.1.2.3 Step 3: Annual Manufacturing Improvement Budgets Development

6.1.2.3.1 Annual Improvement and Cash Budgets for Existing and New Products

To support the annual MCI goal by meeting annual MCI targets at AA-Plant, the MCPD team (in particular, managerial and cost accounting manager, production manager and maintenance manager, quality assurance manager, improvement specialist) has developed *improvement budgets* for PFC3. Some of the development of the improvement budgets for PFC3 was achieved during the previously described establishment of MCI targets and means with MCICP. The mid-year "N" AA-Plant master budget revaluation has considered the improvement budgets and *annual MCI goal* to achieve unit cost reduction by 5.5% in year "N" and then by 6% for year "N + 1" by setting the future performance target of the budgeted cost of products.

Three types of improvement budgets have been developed at AA-Plant for PF3:

1. AMIB for existing products, which aimed to establish MCI targets at each cost structure in each process in the three areas (*processes in area 1, technological flow for injected parts; processes in area 2, technological flow for painted parts; and processes in area 3, assembly line area*). AMIB aimed at continuously monitoring the costs of each process/cost center, both fixed and, especially, variable. In this way, the annual MCI goal of $514,616—the unnecessary cost to be eliminated (or the annual domestic profit to be tackled by productivity gains with kaizen projects in particular)—was detailed at the level of the processes of the three areas as such:
   ■ Manufacturing controllable cost:
     – For equipment losses ($341.492)
     – For human work ($26.231)
     – For materials/energy ($131.154)
     – For waste/stocks ($15.739)
   ■ *Manufacturing uncontrollable cost* (general and administrative costs, logistic costs, financial costs, other costs) did not aim to establish MCI targets and means in the first year. Activities were subsequently established to reduce them.
2. *Multiannual manufacturing improvement budget for new products,* which identified future opportunities to improve the cost of new products, especially in the first three to four months after product launching.

It was determined that for the two new products related to PFC3, which were launched in year "N," the annual MCI goal, after launching the products, was $16,459. This reserve could be exploited by approaching MCI targets and means that was similar to that of old products.

3. *AMCIB* aims to monitor the performance of projects to improve manufacturing costs in terms of generated cash flows. Every 12 months, it has been set to carry out the cash assessment of systematic and systemic improvements implemented to determine the real feasibility of MCI projects.

6.1.2.4 Step 4: Annual Action Plan for MCI Means

Moreover, the annual MCI action plan was developed with the help of MCICP, for the last six months of year "N" and for the first six months of "N + 1." A number of 11 kaizen strategic projects and a kaikaku (replacement equipment) project were defined for the entire manufacturing flow of PFC3 in order to meet the annual MCI targets. The definition of the actions and measures required to implement the solutions to achieve the annual MCI targets was based on the following: (1) MCI, defining/preparing improvements; (2) MCI, solutions, resources, and benefit; (3) MCI, evaluation stage; (4) MCI, gains; and (5) MCI, projects evaluation. At the same time, *individual annual MCI means* were developed.

6.1.2.5 Step 5: Engage the Workforce to Achieve the Annual MCI Targets

The continued and true involvement of all employees in PFC3 and the MCPD team to achieve the annual MCI targets required organization at department level to ensure all necessary resources, especially the time needed to participate in the activities of the MCPD pilot project. The departmental organization focused on *visual management* development for MCPD, structure and duration of *daily MCPD meeting at the shop floor level* (maximum 10 minutes each morning to discuss the five most relevant issues of the previous day regarding the deviations between MCI targets and current stage), *monthly review of KPIs* for CLW for each process, *supporting the work recognition system* of all those involved in the MCPD pilot project, and monitoring the behavior and real involvement of all those involved in the MCPD pilot project. The MCPD team (especially the HR manager and production manager) has developed the training plan for all people involved in reaching the annual MCI targets by the end of year "N" and for year "N + 1" to support the MCI annual action plan and, implicitly, the current and future transformations from processes, through the continuous development of employees' skills and knowledge, and the creation of a flexible learning system. The need to continuously meet the annual MCI targets has led to a good targeting of training needs for all people in PFC3 and beyond. For example, the need to improve the skill level for maintenance department has been identified.

6.1.2.5.1 Systematic and Systemic Improvement Projects (Kaizen and Kaikaku Projects)

To avoid as far as possible the improper and incomplete implementation of the improvement projects foreseen in the annual MCI action plan to achieve the annual MCI targets for PFC3, strict monitoring of all improvement projects, especially of the systematic ones (strategic kaizen), was necessary to change the reactive decision-making managerial behavior into a proactive and preventive one. All activities of the 11 kaizen strategic projects and the systematic improvement project (kaikaku—replacement of equipment to ensure the increase of capacity requested by customer C1 and reduction of operating costs) were presented in the *MCPD Information Centers* highlighting the wide evolution of all activities of the pilot MCPD project (see Case Studies 1, 2, and 3).

6.1.2.6 Step 6: MCI Performance Management

It focused especially on two aspects: (1) confirming the effects of the systematic and systemic improvement projects on the real level of the unit manufacturing cost and on profit (the two objectives of the MCPD pilot project for PFC3) and (2) the contribution of each department to the achievement of the annual MCI targets. At the end of year "N," the level of cost reduction at PFC3 was 5.75% (0.75% above initial target, especially through the achievement of the annual MCI targets, but also through better control of the unit manufacturing costs; some positive effects were noticed subsequently) and the profit was 11.8% (1.8% above the initial target). These results have been achieved by successfully completing the annual MCI targets and achieving the annual external profit from sales/production. Annual MCI targets also aimed at unlocking current capabilities imposed by customers, through strategic kaizen projects and implementing the kaikaku project for the new equipment. *Therefore, the two objectives of the MCPD pilot project at PFC3 have been successfully met.* The planned performance level was met without exceeding the set targets by far and, implicitly, without any effort or allocating too much resources than needed. In this way, the time needed to achieve the production was allocated—annual external profit expected from sales/production. Continuous tracking of the feasibility of all improvement projects was achieved with the *AMCIB*. Some solutions implemented and verified at PFC3 have been transferred to PFC1 and PFC2. The participation rate of PFC3 employees in MCPD activities for the 15 months was 90%.

6.1.2.7 Step 7: Daily MCI Management

Any deviation of the manufacturing costs at the shop floor level for PFC3 and, implicitly, from the annual MCI targets has been addressed through problem-solving techniques (PST), especially using "A3" (annual MCI means level 1—PST). The following were established: all points of consumption monitoring at the three PFC3 areas responsible for each type of consumption, consumption control frequency, and escalation plan in the event of consumption exceeded above the maximum allowable limit. A3 projects were run on a daily basis. In total, for the last six months of the year "N" of the pilot project, 390 A3 projects were carried out and implemented, which were connected to the fulfillment of the annual MCI targets and the annual MCI goal (internal profit), and the average number of internal suggestions was 11/person. At the same time, the behavior of managers (management branding) and of all the people involved in the pilot project of MCPD was monitored continuously to intervene effectively.

6.1.3 Case Studies: Running the Activities and Actions of the Annual MCI Means to Meet Annual MCI Targets

6.1.3.1 Case Study 1: MCI by Improving Equipment Setup and Adjustment Time and Associated Costs (with KAIZENSHIRO)

As specified in step 2 of MCPD, the setup and adjustment loss is 9.18 hours or $771.4 (see Figure 6.6) and resulting WIP from setup loss is 30 pieces or $38.5. At the same time, setup and adjustment loss is a bottleneck activity that causes costs for WIP from setup loss in the "process e," but also over the entire manufacturing flow of PFC3. Total CCLW for PFC3 was calculated at $79,500. For year "N," there have been set annual CCLW targets of $55,000—to contribute to the annual MCI goal of $514,616 (for year "N") at "AA-Plant."

To improve the setup costs and adjustment losses for thermoforming machine (SAE04), the *MDC* will be used (systematic improvement project)—reduce the average setup and adjustment time.

Period of the systematic improvement project (setup and adjustment time) is *between weeks 10 and 15 of year "N."*

The change time is defined as the time from the last piece produced (of previous model *P30*) until the first piece (of the new model *P70F*) produced under series.

To improve the setup costs and adjustment losses for thermoforming machine (SAE04), an interdisciplinary team of specialists from the company was formed together with the MCPD consultant.

The reasons for choosing this systematic improvement project are as follows:

1. Achievement of CLW targets and implicit annual MCI goal
2. Balancing production (fitting to takt time)—increasing MLT synchronization level at DLT
3. Increasing the equipment flexibility (capacity to accept small batches—growing demand from customers, especially from the *C1* customer)
4. Increasing the equipment capacity due to increased customer demand for products made
5. Reducing stocks (raw materials and WIP)
6. Eliminating the perception that new equipment is needed
7. Lack of time for testing new products

Main directions of investigation are as follows: standardization of series change actions, standardization of devices used, the use of quick clamping devices, how to use additional tools, how to perform operations in parallel, layout analysis, the use of mechanical tools, and the 5S level (5S checklist especially for cleaning).

Figure 6.9 presents the basic processes of thermoforming machine (SAE04). *Thermoforming* is a process that uses heat, pressure, and vacuum to form parts of an extruded plastic plate. The plate is heated to a temperature of 165°C to the point where it could become deformed. Then the plate is placed in an aluminum mold. The piece is then cut to the required shape and size. In the last step, the piece is stacked and used further in the manufacturing process.

The systematic improvement project aims at reducing the changing time for aluminum mold from the model "P30" to model "P70F." For this process, a board with thickness between 3.4 and 4.4 mm is used.

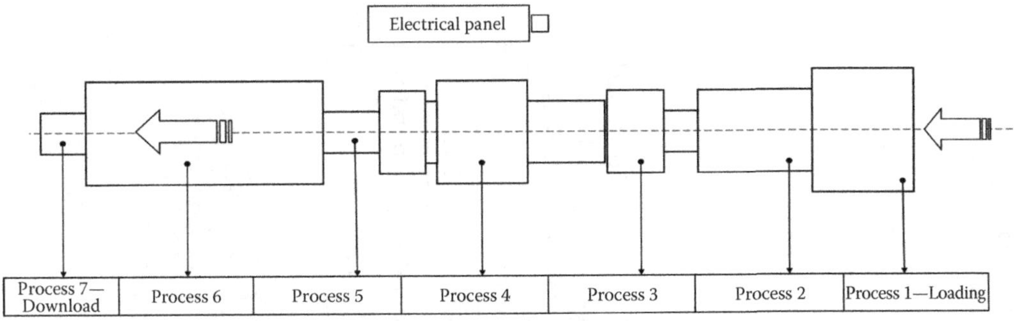

Figure 6.9   The processes for thermoforming machine (SAE04).

Table 6.3 presents the basic parameters of the thermoforming processes (thermoforming machine—SAE04).

The project steps (MDC steps) are as follows (Posteucă and Sakamoto, 2017, pp. 327–352).

6.1.3.1.1 Step 1 of MDC: Setting a Model of Work Contents of a Work Module

This step includes the registration of all initial activities and analyzing input [IP] and output [OP] equipment conditions.

The registration of initial activities was carried out with the aid of video analysis technique in order to understand the current situation. The moment of video analysis was considered representative (on a Tuesday, before lunch, and with average experienced operators). An interdisciplinary team video-recorded all activities during the setup and adjustment time for thermoforming machine (SAE04) (from the last piece of the model "P30" to the first part of the new model "P70F," under series).

Remarks from the video analysis are as follows:

■ Data were collected from the equipment records, which monitors the equipment downtime.
■ Operators are collecting the information—sometimes they make incomplete data records.
■ Production planning is done weekly, then daily; the number of changes are not considered.
■ Setup and adjustment time is included in cycle time in proportion of 15%—regardless of the number of changes.
■ Based on video analysis, data in Table 6.4 *resulted (see 1, II, and III)*.

Table 6.3 Basic Parameters and Information for Thermoforming Machine—SAE04

Process	Loading Plates Area (Process 1)	Preheating Area (Process 2)	Heating Area (Process 3)	Forming Area (Mold Area) (Process 4)	Transfer Area (Process 5)	Cutting Edge Area (Process 6)	Download Area (Process 7)
Time	×	45 seconds	40 seconds	47 seconds	×	35 seconds	×
Temp.	35°C	120°C	165°C	55°C	50°C	40°C	35°C
Operators	×	×	×	×	1 operator	1 operator	1 operator
Operators for setup	×	×	×	×	1 operator	1 operator	×

Table 6.4 Times of Activities for Setup and Adjustment for Thermoforming Machine and KAIZENSHIRO to Achieve CCLW Target for Setup

Input (IP) and Output (OP) Equipment Conditions (Step 1 MDC) (I)	Setup and Adjustment Activities (Step 1 MDC) (II)	No.	Setup and Adjustment Time (from Model "P30" to Model "P70F")						BF/ AF (IV)	E	C	R	S
			Op. 1 Sec (III)	Op. 2 Sec (III)	Op. 3 Sec (III)	Time Start (III)	End (III)	Duration Hours:Min: Sec (III)		(V)			
Automatic	Stopping the loading of plates	1.	10	0	0	00:00:00	00:00:10	00:00:10	AF	×			
	Stopping the heating of the mold	2.	30	0	0	00:00:10	00:00:40	00:00:30	AF	×			
	Preparing tools and production program; waiting for the loading of 3 plastic plates	3.	123	0	0	00:00:40	00:02:41	00:02:01	AF	×	×		×
	Bursting air	4.	40	0	0	00:02:41	00:03:21	00:00:40	AF			×	
	Stopping the production cycle (model "P30")	5.	60	0	0	00:03:21	00:04:21	00:01:00	BF			×	

(Continued)

Table 6.4 (Continued) Times of Activities for Setup and Adjustment for Thermoforming Machine and KAIZENSHIRO to Achieve CCLW Target for Setup

Input (IP) and Output (OP) Equipment Conditions (Step 1 MDC) (I)	Setup and Adjustment Activities (Step 1 MDC) (II)	No.	Op. 1 Sec (III)	Op. 2 Sec (III)	Op. 3 Sec (III)	Time Start (III)	Time End (III)	Duration Hours:Min: Sec (III)	BF/ AF (IV)	E	C	R	S
The last piece of the model "P30"													
Manual	Opening the door of the mold	6.	20	0	0	00:04:21	00:04:41	00:00:20	AF				
	Bringing the flanges and changing the last of the guillotine	7.	0	60	0	00:04:41	00:05:41	00:01:00	AF			×	×
	Starting the cooling of the mold and detaching the mold guide	8.	30	0	0	00:05:41	00:06:11	00:00:30	AF			×	×
	Setting the heating plate	9.	50	0	0	00:06:11	00:07:01	00:00:50	AF				
	Unlocking and lowering the mold	10.	20	0	0	00:07:01	00:07:21	00:00:20	AF			×	×
	Closing water valves	11.	30	0	0	00:07:21	00:07:51	00:00:30	AF			×	×

(Continued)

Table 6.4 (*Continued*)   Times of Activities for Setup and Adjustment for Thermoforming Machine and KAIZENSHIRO to Achieve CCLW Target for Setup

Input (IP) and Output (OP) Equipment Conditions (Step 1 MDC) (I)	Setup and Adjustment Time (from Model "P30" to Model "P70F")												
	Setup and Adjustment Activities (Step 1 MDC) (II)	No.	Op. 1 Sec (III)	Op. 2 Sec (III)	Op. 3 Sec (III)	Time Start (III)	End (III)	Duration Hours:Min: Sec (III)	BF/ AF (IV)	E (V)	C	R	S
	Clamping of mold bracket, hose uncoupling, and extraction of the mold (for model "P30")	12.	50	0	0	00:07:51	00:08:41	00:00:50	BF			×	×
	Moving to rack dies (for model "P30")	13.	50	0	0	00:08:41	00:09:31	00:00:50	BF	×	×	×	×
	Setting the feeder plates	14.	0	20	0	00:09:31	00:09:51	00:00:20	AF	×	×	×	
	Protecting the upper column and setting feeder suction cups	15.	0	30	0	00:09:51	00:10:21	00:00:30	AF	×	×	×	
	Positioning the mold on the rack	16.	40	0	0	00:10:21	00:11:01	00:00:40	AF			×	×
	Receiving the new mold (for model "P70F")	17.	70	0	0	00:11:01	00:12:17.	00:01:16	AF		×	×	×

(Continued)

Table 6.4 (Continued) Times of Activities for Setup and Adjustment for Thermoforming Machine and KAIZENSHIRO to Achieve CCLW Target for Setup

Input (IP) and Output (OP) Equipment Conditions (Step 1 MDC) (I)	No.	Setup and Adjustment Activities (Step 1 MDC) (II)	Setup and Adjustment Time (from Model "P30" to Model "P70F")			Time		Duration	BF/ AF (IV)	E	C	R (V)	S
			Op. 1 Sec (III)	Op. 2 Sec (III)	Op. 3 Sec (III)	Start (III)	End (III)	Hours:Min: Sec (III)					
	18.	Traveling with new mold for thermoforming machine	80	0	0	00:12:17	00:13:50	00:01:33	AF			×	×
	19.	Positioning new mold—ring (for model "P70F")	60	0	0	00:13:50	00:14:50	00:01:00	BF			×	×
	20.	Inserting new mold—start warming (for model "P70F")	50	0	0	00:14:50	00:15:40	00:00:50	BF			×	×
	21.	Moving to the control panel/installing electrical couplings	10	0	0	00:15:40	00:15:50	00:00:10	AF				
	22.	Entering the parameters (for model "P70F")	20	0	0	00:15:50	00:16:10	00:00:20	AF	×			
	23.	Checking/idling and then start of feeding with plastic plates	20	0	0	00:16:10	00:16:30	00:00:20	AF			×	×

(Continued)

Table 6.4 (Continued)   Times of Activities for Setup and Adjustment for Thermoforming Machine and KAIZENSHIRO to Achieve CCLW Target for Setup

Input (IP) and Output (OP) Equipment Conditions (Step 1 MDC) (I) No.	Setup and Adjustment Activities (Step 1 MDC) (II)	Op. 1 Sec (III)	Op. 2 Sec (III)	Op. 3 Sec (III)	Start (III)	End (III)	Duration Hours:Min: Sec (III)	BF/ AF (IV)	E	C	R	S
										(V)		
24.	Sampling/adjust the cutting block	20	0	0	00:16:30	00:16:50	00:00:20	AF	×		×	×
25.	Checking and blocking the frame of the mold	10	0	0	00:16:50	00:17:00	00:00:10	AF				
26.	Tightening the guides for the mold	20	0	0	00:17:00	00:17:20	00:00:20	AF			×	×
27.	Closing the doors of the mold (for model "P70F")	20	0	0	00:17:20	00:17:40	00:00:20	BF			×	×
28.	Air coupling	10	0	0	00:17:40	00:17:50	00:00:10	AF				
29.	Heating the mold (for model "P70F")	480	0	0	00:17:50	00:25:50	00:08:00	BF	×			
30.	Starting the automatic cycle	10	0	0	00:25:50	00:26:00	00:00:10	AF				

(Continued)

Table 6.4 (*Continued*) Times of Activities for Setup and Adjustment for Thermoforming Machine and KAIZENSHIRO to Achieve CCLW Target for Setup

Input (IP) and Output (OP) Equipment Conditions (Step 1 MDC) (I)	No.	Setup and Adjustment Activities (Step 1 MDC) (II)	Op. 1 Sec (III)	Op. 2 Sec (III)	Op. 3 Sec (III)	Time Start (III)	Time End (III)	Duration Hours:Min:Sec (III)	BF/AF (IV)	E	C	R (V)	S
Automatic	31.	Adjusting fans and timer	20	0	0	00:26:00	00:26:20	00:00:20	AF			×	×
	32.	Moving behind the equipment	30	0	0	00:26:20	00:26:50	00:00:30	AF			×	×
	33.	Opening water valves	20	0	0	00:26:50	00:27:10	00:00:20	AF			×	×
	34.	Adjusting the pressure	20	0	0	00:27:10	00:27:30	00:00:20	AF				
	35.	Checking the thermoforming of the first parts (for model "P70F")	20	0	0	00:27:30	00:27:50	00:00:20	AF			×	
	36.	Moving to the control panel	20	0	0	00:27:50	00:28:10	00:00:20	AF	×			
	37.	Adjusting the parameters	20	0	0	00:28:10	00:28:30	00:00:20	AF	×	×		
	38.	Checking the first part of the new model (completed)	30	0	0	00:28:30	00:29:00	00:00:30	AF			×	×

Abbreviations: Elimination, E; Combination, C; Rearrangement, R; Simplification, S.

6.1.3.1.2 Step 2 of MDC: Defining Functions of All Work Contents

This step includes the identification of basic function (BF) and auxiliary function (AF). Following the video analysis, the following *basic functions* of setup and adjustment activities have been identified for thermoforming machine (SAE04) (see Table 6.4 IV):

■ Stop production cycle (model "P30")
■ Mold clamping bracket, hose uncoupling, and extraction mold (for "P30")
■ Moving to rack dies (for model "P30")
■ Positioning new mold—ring (for model "P70F")
■ Insert new mold—start warming (for model "P70F")
■ The mold closing doors (for model "P70F")
■ Heating mold (for model "P70F")

6.1.3.1.3 Step 3 of MDC: Setting Design Target as Improvement Value

This step includes the improvement MCI means target or KAIZENSHIRO to meet CCLW target. The setup and adjustment time improvement target was set at nine minutes—MCI means target (from the current 29 minutes) to meet the annual CCLW target (implicitly annual MCI targets, by participating CCLW for setup activity to accomplish annual MCI goal). The target number of setup and adjustment per month is 30 (providing the average number required by customers).

Based on the *spaghetti diagram*, it was found that

■ Operator/setter 1 (op. 1) moves for 162 meters during the setup and adjustment activities and performs most operations.
■ Operator/setter 2 (op. 2) moves for 18 meters during the setup and adjustment activities.
■ Operator 3 (op. 3) does not work on setup and adjustment activities.
■ Movement for storage of the old mold ("P30" model): 17 meters (operator/setter 1).
■ Movements for transferring the new mold ("P70F" model): 72 meters (operations 13, 16, 17, and 18; Table 6.4) (operator/setter 1).
■ Transfer time for molds is 240 seconds or 4 minutes: Operation 13–50 seconds, operation 16–40 seconds, operation 17–70 seconds, and operation 18–80 seconds (Table 6.4).

Other remarks on the current state are as follows: mold heating medium is water; working temperature is 80°C; mold heating time is 480 seconds or 8 minutes; operators expect 390 seconds or 6.5 minutes.

Main problems identified are as follows:

■ *Problem 1*: Long time to insert new mold 110 seconds (operation 19–60 seconds and operation 20–50 seconds)
■ *Problem 2*: Long time to heat the mold (operation 29–480 seconds)
■ *Problem 3*: Long time to transfer the molds (operation 13, 16, 17, and 18)
■ *Problem 4*: Operator 1 performs most operations (operator 1, 1123 seconds; operator 2, 110 seconds; operator 3, does not work on setup and adjustment activities)
■ *Problem 5*: Setting parameters for new mold ("P70F" model) (operation 22 and 37)

6.1.3.1.4 Step 4 of MDC: Searching/Creating Improving Ideas

Solutions for the main problems identified by using ECRS (Elimination, E; Combination, C; Rearrangement, R; Simplification, S) (see Table 6.4 V) were determined by asking the following questions related to the current method of work: *Can it be eliminated? Can it decrease? Can it change? Can it be done simultaneously? Can it speed up? Can it be easier? Can it be done elsewhere?*

The solutions identified were as follows:

■ *Solution to Problem 1*: Mounting the new mold with extractors before changing; initial time of 110 seconds has been reduced to 30 seconds.
■ *Solution to Problem 2*: External circuit for mold preheating at the temperature of 90°C. The mold heating time will be done out of the setup and adjustment time; Elimination of the 480 seconds' time.
■ *Solution to Problem 3*: Making a device for the temporary storage of the old mold (for "P30" mold); positioning the preheating device for new mold ("P70F" model) near the thermoforming machine (SAE04). This way, the movement for old mold storage has been reduced from 17 to 3.5 meters, and the movement for transferring the new mold has been reduced from 72 to 12 meters. The time for the initial transfer of 240 seconds has been reduced to 50 seconds.
■ *Solution to Problem 4*: Reorganizing the operators' operations and the setup and adjustment operations (operator 1, 311 seconds; operator 2, 182 seconds; operator 3, 40 seconds); operators' movements (operator 1,

reducing the browsed distance from 162 to 38 meters; operator 2, reducing the browsed distance from 18 to 12 meters; and operator 3, has to move for 9 meters).

■ *Solution to Problem 5*: Presetting the new mold parameters ("P70F" model) (elimination of the operation 22–20 seconds and operation 37–20 seconds).

6.1.3.1.5 Step 5 of MDC: Modifying and Summarizing Ideas as a Concrete New Method

Table 6.5 presents the new Standard Operating Procedures (SOP) for setup and adjustment time for thermoforming machine, to achieve CCLW target (MCI means target—9 minutes).

Based on tests performed, a new work procedure was issued for changing the molds. So, the new diagram for setup and adjustment time (Table 6.5) has become *work procedure*. Based on the new setup and adjustment chart, a *training plan* was designed to help implementing the solutions found for two other similar equipment (horizontal application).

Figure 6.10 presents the evolution in time of setup and adjustment. Time improvement was 71%. The MCI means target (9.00 minutes target) was achieved and exceeded (8.43 minutes).

During analyses, other opportunities for improvement were identified. The main opportunities were regarding the *review of OEE data collection and correlation of production planning with setup and adjustment times*.

6.1.3.1.6 Analysis of Performance Achieved Over Time

The target for the setup and adjustment number of 30 per month was reached and exceeded. It reached 35 setup and adjustments per month.

The systematic improvement costs (MDC solution cost) related to implementing solutions for systematically improving the setup and adjustment time were as follows:

■ $250 cost of materials—purchasing the extractors (CM1)
■ $150 cost of materials—trolley for temporary storage of old mold (for "P30" model) (CM2)
■ $150 cost of trolley supplier for temporary storage of old mold (for "P30" model) (CS)

Table 6.5 New Diagram for Setup and Adjustment Time for Thermoforming Machine (New SOP)—Achieving CCLW Target (for Setup)

Equipment Conditions	No.	Activities	Op. 1	Op. 2	Op. 3	Time		Duration	E/I
			Sec	Sec	Sec	Start	End	Hours:Min:Sec	
Automatic/ preparation setup	1.	Stopping the loading of plastic plates and of heating the mold	5	5	0	0:00:00	00:00:10	00:00:10	E
	2.	Setting the feeder plates	0	17	0	0:00:10	00:00:27	00:00:17	E
	3.	Setting feeder suction cups	0	17	0	0:00:27	00:00:44	00:00:17	E
	4.	Setting the heating plate	15	0	0	0:00:44	00:00:59	00:00:15	I
	5.	Bursting air	15	0	0	0:00:59	00:01:14	00:00:15	I
	6.	Stopping the production cycle (model P30)	18	0	0	0:01:14	00:01:32	00:00:18	I

The last piece of the model "P30"

Equipment Conditions	No.	Activities	Op. 1	Op. 2	Op. 3	Time		Duration	E/I
			Sec	Sec	Sec	Start	End	Hours:Min:Sec	
Manual	7.	Opening the door of the mold	8	8	0	0:01:32	00:01:48	00:00:16	I
	8.	Bringing the flanges and changing the last of the guillotine	0	0	25	0:01:48	00:02:13	00:00:25	I
	9.	Starting the cooling of the mold and detaching the mold guide	0	15	0	0:02:13	00:02:28	00:00:15	I
	10.	Unlocking and lowering the mold	15	0	0	0:02:28	00:02:43	00:00:15	I
	11.	Closing water valves	0	15	0	0:02:43	00:02:58	00:00:15	I

New Diagram for Setup and Adjustment Time (from Model "P30" to Model "P70F")—Achieving CCLW Target (for Setup)

(Continued)

Table 6.5 (*Continued*) New Diagram for Setup and Adjustment Time for Thermoforming Machine (New SOP)—Achieving CCLW Target (for Setup)

New Diagram for Setup and Adjustment Time (from Model "P30" to Model "P70F")—Achieving CCLW Target (for Setup)

Equipment Conditions	No.	Activities	Op. 1 Sec	Op. 2 Sec	Op. 3 Sec	Time Start	Time End	Duration Hours:Min:Sec	E/I
	12.	Clamping the mold bracket, hose uncoupling, and extraction of the mold (for "P30")	15	15	15	0:02:58	00:03:43	00:00:45	I
	13.	Moving to support the temporary storage	10	0	0	0:03:43	00:03:53	00:00:10	I
	14.	Temporary storage (mobile support)	10	0	0	0:03:53	00:04:03	00:00:10	I
	15.	Receiving the new mold (for model "P70F")	15	0	0	0:04:03	00:04:18	00:00:15	I
	16.	Traveling with new mold to the thermoforming machine	15	0	0	0:04:18	00:04:33	00:00:15	I
	17.	Positioning new mold—ring	15	15	0	0:04:33	00:05:03	00:00:30	I
	18.	Inserting new mold—start warming	15	15	0	0:05:03	00:05:33	00:00:30	I
	19.	Checking and blocking the frame of the mold	10	0	0	0:05:33	00:05:43	00:00:10	I
	20.	Moving to the control panel/installing electrical couplings	10	0	0	0:05:43	00:05:53	00:00:10	I
	21.	Tightening guides of the mold	10	0	0	0:05:53	00:06:03	00:00:10	I

(*Continued*)

Table 6.5 (*Continued*)   New Diagram for Setup and Adjustment Time for Thermoforming Machine (New SOP)—Achieving CCLW Target (for Setup)

New Diagram for Setup and Adjustment Time (from Model "P30" to Model "P70F")—Achieving CCLW Target (for Setup)

Equipment Conditions	No.	Activities	Op. 1 Sec	Op. 2 Sec	Op. 3 Sec	Time Start	Time End	Duration Hours:Min:Sec	E/I
	22.	Closing the doors of the mold	10	0	0	0:06:03	00:06:13	00:00:10	I
	23.	Checking/idling and then start feeding with plastic plates	10	0	0	0:06:13	00:06:23	00:00:10	I
	24.	Sampling/adjusting the cutting block	15	0	0	0:06:23	00:06:38	00:00:15	I
	25.	Air coupling	10	0	0	0:06:38	00:06:48	00:00:10	I
	26.	Starting the automatic cycle	10	0	0	0:06:48	00:06:58	00:00:10	I
Automatic	27.	Adjusting fans and timer	15	0	0	0:06:58	00:07:13	00:00:15	I
	28.	Moving behind the equipment	0	15	0	0:07:13	00:07:28	00:00:15	I
	29.	Opening water valves	0	15	0	0:07:28	00:07:43	00:00:15	I
	30.	Adjusting the pressure	0	20	0	0:07:43	00:08:03	00:00:20	I
	31.	Checking the thermoforming of the first parts (for model "P70F")	15	0	0	0:08:03	00:08:18	00:00:15	I
	32.	Moving to the control panel	10	0	0	0:08:18	00:08:28	00:00:10	I
	33.	Checking the first part of the new model (completed)	15	10	0	0:08:28	00:08:43	00:00:15	I

Notes: E, external setup (thermoforming machine is in operation); I, internal setup (thermoforming machine is stopped).

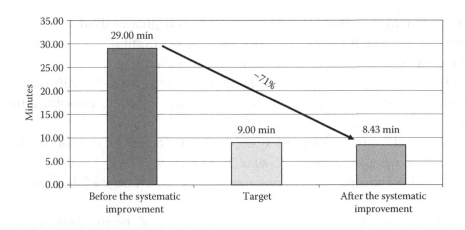

Figure 6.10   The result of improving the setup and adjustment time (annual MCI means targets for thermoforming machine) with MDC.

- ■ $4,200 cost of new mold preheating device ("P70F" model) next to the thermoforming machine (SAE04), including the acquisition of the necessary hoses and quick couplers (CD)
- ■ $597 cost with personnel (PC)—directly productive staff = 10 man*hour/project; indirectly productive staff = 75 man*hour/project; hourly cost for direct labor = $3.45/hour; hourly cost for indirect labor = $7.5/hour

Total costs (C_{tot}) were as follows:

$$C_{tot} = CM1 + CM2 + CS + CD + PC$$

$$= \$250 + \$150 + \$150 + \$4.200 + \$597 = \$5.347 \tag{6.1}$$

Annual CCLW targets of $55,000 have been successfully completed.

6.1.3.2 Case Study 2: MCI by Increasing Productivity with Replacement of Bottleneck Equipment (with KAIKAKU)

An annual MCI action plan at "AA-Plant" has established a kaikaku project (MCI means level 3) to reduce CCLW caused by lack of capacity. The annual value of CCLW was calculated at $78,500.

After analysis during the setup and adjustment time for thermoforming machine (SAE04), the need to increase capacity by replacing the equipment (SAE04) was emphasized.

Furthermore, the market analysis regarding future volumes of products made by the thermoforming process predicts significant growth for the PFC3.

So, the capacity needs for thermoforming are far greater than what can be achieved through systematic improvement of thermoforming machine (SAE04) or other similar equipment in the "AA-Plant."

In the following, one may see the analysis of the need to shift from systematic improvement to systemic improvement of the thermoforming process (thermoforming machine—SAE02).

The period for the project of systemic improvement of the thermoforming process lasted *between weeks 16 and 33 of "N."*

The forecast for the necessary capacities for year ("N + 1") is presented in Table 6.6.

As shown in Table 6.6, since May, the capacity of the thermoforming machine (SAE04) becomes negative. In this context, it is necessary to analyze the opportunity of equipment changing.

Monthly average forecast of need for capacity for thermoforming machine (SAE04) for year "N + 1" is of 16.500 pieces. Therefore, further systematic improvement of the thermoforming machine (SAE04) becomes unnecessary to meet capacity requirements.

Following the strategic decision to change equipment SAE04, an interdisciplinary team of specialists from the company together with the MCPD consultant was formed to analyze the technical needs of the future thermoforming machine. The project team for the systemic improvement of thermoforming equipment was made up of specialists in the following departments: production engineering, supply, finance, product development, marketing planning, maintenance, and quality.

After analyzing the previous experience with the thermoforming machine SAE04 activity and depending on future market needs, the technical needs of the future equipment for each area of it were determined—new thermoforming machine (SAE02).

The steps of the systemic improvement project were as follows:

1. *Concept planning—17 days*

   So, on the monthly capacity needs, on the budget established for systemic improvement (equipment changing), the required equipment capacity (cycle time) and basic features were set. The capacity requirements of the new equipment were determined based on simulation to an OEE level of 0.93 (OEE = VAOT/ELT = 12,000/12,900 = 0.93 or *availability of 0.94, performance of 0.99*, and *quality of 1.00*), with a standard cycle time of 39 seconds (connected to takt time). The monthly necessary production capacity is of 18.650 pieces.

Table 6.6 Analysis of Required Production Capacity for the Thermoforming Machine (SAE04) in "N + 1"

	Jan.	Feb.	Mar.	Apr.	May	June	July	Aug.	Sep.	Oct.	Nov.	Dec.
Required monthly production	11.577	11.977	12.077	12.177	12.527	12.927	12.800	12.900	13.000	13.150	13.300	13.500
Normal capacity	11.800	11.800	11.800	11.800	11.800	11.800	11.800	11.800	11.800	11.800	11.800	11.800
Capacity over time	0	250	300	450	450	450	450	450	450	450	450	450
Total capacity	11.800	12.050	12.100	12.250	12.250	12.250	12.250	12.250	12.250	12.250	12.250	12.250
Capacity difference	223	73	23	73	−277	−677	−550	−650	−750	−900	−1.050	−1.250

The budget set for this new equipment was \$635,000. This budget was set in when determining the life cycle of the products that were to be made with this equipment. The fixed cost of depreciation of equipment is part of the unit manufacturing cost of products and cannot be exceeded.

The basic characteristics have been determined by the *production engineering department* for reduction size, height of thermoformed part, automatic feeding, automatic cross of plastic part, preheat of plastic parts, hydraulic table, guillotine sizes, hourly productivity levels, part evacuation system, and air pressure.

The basic principles followed by the *production engineering department* for future equipment (new thermoforming machine—SAE02) were at the level of main losses and waste that may occur in the future:

■ *Reliability*: Causes that may affect the availability target of 0.94 were considered the potential ways of decreasing the hourly productivity for SAE02 (other thermoforming equipment). So, the potential causes of failure, the potential causes of small stops, the ease of adjustment, and the stability of cycle time were analyzed.

■ *Maintainability*: The degree of ease to detect operating parameters that do not comply and ease of maintenance interventions was analyzed.

■ *Operability*: The needed characteristics of future equipment to ensure a safe and easy operating from the operators were analyzed.

■ *Quality*: The possible level of scrap and rework was analyzed.

■ *Flexibility*: The necessary time for setup operations was analyzed.

■ *Energy consumption*: The electricity consumption was analyzed.

The finance department together with managerial accounting department made the following financial projections: the standard unit cost calculation, the corresponding operating budget, and the investment budget.

Following this analysis, three main conclusions resulted that

■ They no longer continue with the current thermoforming machine (SAE04).

■ They cannot achieve such equipment within the company.

■ They need to look for a provider that fulfills the necessary characteristics of SAE02.

2. *Action plan and design—35 days*

After determining the equipment features, the mold specifications (size and shape of the interface with the new thermoforming machine), the cost of commissioning, the level of the future operational costs, the level of CLW on the equipment, and the flexibility level (15 minutes initially setup, mold preheating made by the equipment, and rapid evacuation of the mold) were set.

They then analyzed the offers received from suppliers. The assessment criteria for offers were price, cost of transport, cost of commissioning, life cycle cost analysis for new equipment, delivery time, setup time, payment terms, lead time for spare parts, and process capability (Cp, Cpk).

Four offers were received (the name of suppliers: "C," "K," "N," and "B"). Following the analysis, the supplier "K" was selected.

The "K" supplier has designed a new thermoforming machine with a monitoring system for plate temperature and a transfer system for ball plates (according to the requirements resulting from historical issues of SAE04).

3. *Fabrication—30 days*

The equipment manufacturing and testing in accordance with the requirements lasted for 30 days. A team of two specialists from the company paid two visits to the manufacturer's factory to perform the verification of the equipment design and to make the requested design changes.

4. *Transport and installing of new equipment—17 days*

The transport of new equipment lasted for seven days. The installation was performed by the manufacturer and lasted for 10 days. Prior to the installation, layout tests were conducted and SAE04 was uninstalled. Uninstalling SAE04 was done by company employees.

5. *Testing of products and start of production—20 days*

Product testing and operator training lasted 20 days.

6.1.3.2.1 Analysis of Performance Achieved Over Time

The resulting cycle time was 39 seconds (the target was met). The initial OEE was 93% (the target was met). The defect rate was zero. The total cost of systemic improvement (the investment in new equipment) was $604,000 (the target was met). Production start time was 119 days or 3.9 months (target was 3.6 months; there was a delay of 11 days).

The expected yearly benefit of the new equipment is $693,000/year (16,500 pieces per month * 12 months * $3.5 per piece).

Return on investment (ROI) for new equipment

$$ROI = Total\ revenue - Investment\ cost/Investment\ cost * 100$$

$$ROI = \$693,000 - \$604,000 / \$604,000 * 100 = 14.73$$

The ROI result of 14.73 is considered good because the company believes that 12 is the minimum necessary.

Annual CCLW targets have been successfully reached ($78,500).

6.1.4 Conclusion

Therefore, the MCPD team has established the annual MCI targets and means to meet the two objectives of the MCPD system pilot project (minimum 5% reduction of unit manufacturing costs per year and minimum 10% annual target profit from capacity increase based on customers' rising demand—external profit and, especially from meeting the annual MCI goal, internal profit). The annual MCPD stake for the annual MCI goal was about $500,000, or, in other words, the annual target revenue is $500,000 (equivalent to *total annual CLW targets* for all related PFC3 processes). The processes that are relevant to customers who place the greatest pressure on unit manufacturing costs reduction take precedence in establishing the annual MCI targets, and the annual CCLW targets take precedence in establishing the annual MCI means targets.

The annual MCI assimilated goal is also the profit obtained from the improvement of the cost of raw materials, auxiliary materials, and utilities, in particular by reducing unnecessary consumption and by negotiating prices with suppliers and/or by designing or redesigning products and/or equipment. *Material cost reductions* through successive price negotiations with suppliers (*supply chain management area*) and product design and redesign (*product design area*) are not included particularly in the MCPD system. At the end of the MCPD pilot project at PFC3, the MCPD team has turned into a distinctive MCPD department responsible for coordinating internal profitability through the achievement of the annual MCPD (*MCPD steering committee*).

The greatest operational challenge of the MCPD system pilot project at AA-Plant was in the *Manufacturing Cost Policy Development* phase because, even though the annual MCI targets and means have been properly localized and dimensioned in advance, in the *Manufacturing Cost Policy Analysis* phase, a relative state of "relaxation" of the MCPD team emerged. This "relaxation" had several causes, but three were the main ones:

1. Awareness of the actual level of CLW for the area of each manager caused a certain fear of proposing convergent improvement activities and actions to meet the annual MCI goal for the last 6 months of year "N"

2. Slightly exaggerated anchoring in the operational activities of the MCPD team to seek some excuses regarding the delay of proposing the annual MCI means targets to achieve the MCI targets
3. Proposing insufficient initiatives to fully reach the annual MCI targets

The attitude and contextual behavioral identity of the plant manager at AA-Plant were essential to mitigate and then eliminate these temporary operational challenges.

The greatest strategic challenge of the MCPD system pilot project at AA-Plant was in the *Manufacturing Cost Policy Development* phase and aimed at the plant manager assuming a business scenario for the future to meet the annual MCI targets and means. As the turnover trend for PFC3 was slightly rising, the scenario for achieving the annual MCI targets was to allocate 65% of the annual MCI goal burden on cost reduction, reducing variable costs (especially indirect), and 35% for capacity increase (by volume).

Following the pilot project of the MCPD system, both the plant manager and the entire MCPD team have become aware that the MCPD system's stake is to continuously ensure the *uncovering of the hidden reserves of profitability* by continuously fulfilling the annual MCI targets and means in a scientific, systematic, and systemic manner, continuously realizing the connections between losses, waste, costs, and processes at the level of each PFC and on the overall company.

6.2 "BB-Plant": Process Industry

6.2.1 Establishment of MCPD System: Actions, Activities, and Challenges

The following is a case study of implementing the MCPD system at BB-Plant. The approach was roughly similar to "AA-Plant." Two MCI improvement projects will be presented.

BB-Plant is a multinational firm operating in the food industry. AA-Plant is one of the top three companies in the group, with approximately 350 employees. The production regime is one in lots. Following the initial analyses, two *product family costs (PFC)* (PFC1 and PFC2—75 employees) have been identified.

The main objective of the plant manager was to increase profitability and to identify ways to reduce the manufacturing cost for the coming years and to increase the number of consecutive production days without stopping, to increase OEE, to support quality, to improve efficiency of shutdown, and to reduce power consumption.

For this, a MCPD implementation team called "Manufacturing System Transformation by MCPD" was formed.

Implementation targeted PFC2 and lasted 12 months.

The following main CLW structures have been identified for the following: (1) *operation losses*, production adjustment, quality defect, reprocessing; (2) *equipment losses*, shutdown, production adjustment, equipment failure, process failure, normal production, abnormal production, quality defect, reprocessing; (3) *material/yield and energy losses*, production adjustment; and (4) *waste* (flow, inventory, WIP), production adjustment.

The annual MCI goal for BB-Plant was $1,260,000. The annual MCI goal for PFC2 was $850,000. The annual MCI targets and means for the eight manufacturing phases have been set. The annual improvement budgets for PFC1 and PFC2, the annual MCI action plan (12 kaizen and 1 kaikaku projects), the departmental organization, and annual MCI training plan have been developed. We will present one of the 12 kaizen projects (annual strategic) for improving operator skills to sustain quality and one of the 27 A3 projects deployed in the first year of implementing the MCPD system (cost problem solving with the lubricants consumption for one equipment). The average number of annual internal suggestions was 8/person. The *MCPD information center* presented the evolution of all *annual MCI means*.

6.2.2 Case Studies: Running the Activities and Actions of the Annual MCI Means to Meet Annual MCI Targets

6.2.2.1 Case Study 1: MCI by Improving Operators' Skills to Sustain Quality (with KAIZEN)

The objective of the kaizen project was to reduce the CLW related to quality defect and reprocessing for PFC2 by improving operator skills for task "x" and restoring the work method. The annual CLW target was $84,500.

PLAN—Location, Evaluation, and Objectives
1. *Locating the need for improvement:*
   ■ The quality deficiencies (quality defect and reprocessing) were identified from all areas and these were quantified in CLW: two areas in the eight phases of the production related to PFC2.
   ■ The used materials were identified.

- The current work method was analyzed.
- Two ways of performing task "x" ("x1" and "x2") were identified.

2. *Assessing the current performances of operators*:
   - There were identified three workstations where quality-related CLWs and a number of nine operators performing the task "x" (for each of the two ways of performing) were identified.
   - Data on the performance related to quality of the nine operators for the last six months was collected: operator 1 ("x1"), 0.12%; operator 2 ("x1"), 0.9%; operator 3 ("x1"), 0.7%; operator 4 ("x2"), 0.6%; operator 5 ("x2"), 0.9%; operator 6 ("x2"), 0.11%; operator 7 ("x1" and "x2"), 0.16%; operator 8 ("x1" and "x2"), 0.14%; operator 9 ("x1" and "x2"), 0.12% (the fluctuating performance of operators has determined the need for training).

3. *Necessary training and budget identification*:
   - Seven principles for performing "x1" and "x2" were defined.
   - Four principles contributing to the occurrence of quality problems were identified.
   - Four abilities needed to be improved by the nine operators were identified to perform "x1" and "x2" by observing the four partially respected principles.
   - The current matrix skills of operators for "x1" and "x2" were analyzed regarding the four types of deficient skills of the nine operators; each operator was assessed: (1—beginner) for beginner operators with good theoretical knowledge; (2—medium) for operators with good theoretical knowledge, but less practical; (3—experienced) for operators with good theoretical and practical knowledge; and (4—expert) for operators who can teach others; the average of the initial skills of the nine operators, for the four abilities necessary to be improved to support the required quality level and to meet the CLW target—in terms of quality and not only, was 2.77.

4. The increase of the average skills of the nine operators from 2.77 to 3.5 was established.

5. The training needs were established:
   - Six hours' theoretical training for a group of nine operators.
   - Four hours' workshop for a group of nine operators.

- Planning at the training room, projector, school supplies, teaching materials.
- A training map specific to the four abilities required to be improved was prepared in order to support the necessary quality level (prepared by the following departments: quality, maintenance, and industrial engineering).
- A film was prepared to explain the current state and the future state required for the tasks "x1" and "x2"—the four skills required to be acquired.
- The training and the workshop will be carried out with an internal trainer—a specialist in the maintenance department (outside the program hours).
- Training method: video presentation, explanation, group discussions, and individual reading.
- Evaluation of the training program: written test, interview, and feedback questionnaire.
- Workshop method: practical demonstration, case studies, and questions.
- Evaluation of the workshop: practical test and simulations.

6. *Establish the training program objectives*: by setting the evaluation standards for operators to increase their abilities (from 2.77 to 3.5) the increase of the quality of products was established, by decreasing the quality defect and reprocessing, at <499 ppm or <0.05%, meeting the annual MCI means targets (from 2.77 to 3.5) in order to meet the annual CLW target for quality of $84,500.

DO: Training Program

7. Supporting the theoretical course and the workshop; the theoretical and practical explanation of incorrect and correct activities.

CHECK: Training Evaluation

8. The theoretical knowledge gained through training was verified—a grid of 20 theoretical questions.
9. The training program was evaluated (with grades from 1 to 10).
10. Each operator was practically assessed.
11. An average grade was made with the grades from the theoretical and practical evaluation.
12. Skills evaluation after training and workshop:
    The average of capabilities before was Ppk of 1.24 (Ppk = Process Performance).

The average of capabilities after was Ppk of 1.49 (Ppk = Process Performance).

13. The operators with capability <1.33 are repeating the training program (three operators).

14. After completing the training program, the matrix skills of the nine operators for "x1" and "x2" regarding the four types of abilities necessary to be improved reached an average of 3.5 (annual MCI means target was met).

15. Annual MCI target (annual CLW target for quality) of $84,500 was met.

ACT: Training Plan for the Future

16. SOP were adjusted.

17. The identified improvements could not be extended to other activities; three other types of deficiencies in the tasks performed by the nine operators were identified and a plan for their approach was made.

6.2.2.2 Case Study 3: Daily MCI Management—Cost Problem Solving with the Consumption of Lubricants for One of the Equipment (with A3)

Daily MCI management is based on the Plan-Do-Check-Act (PDCA) cycle. At the shop floor level, the PDCA cycle is approached using problem-solving techniques, such as A3 (annual MCI means level 3—PST).

One of the issues analyzed at the shop floor level was the lubricants consumption improvement for "YY" equipment. This consumption increased from 1,550 grams each month (current standard) to 1,850 grams (monthly average), in the last four months. This information was gathered from SAP—costs module (cost center for "YY" equipment). The new target was set at 1,400 grams (the future standard of lubricant consumption on the equipment; MCI means targets level 1—PST). Therefore, the purpose of MCI through MCPD is to investigate the difference between the current state and the target set and to choose the means to reach the target within the set time frame (not to analyze the variations between the current state and the established standard; continuous and consistent fulfillment of improvement targets cost is more important than maintaining a current standard cost).

The period for the improvement project was *between weeks 20 and 26 of N.*

Figure 6.11 presents how to address the problem of exceeded lubricants consumption for "YY" equipment.

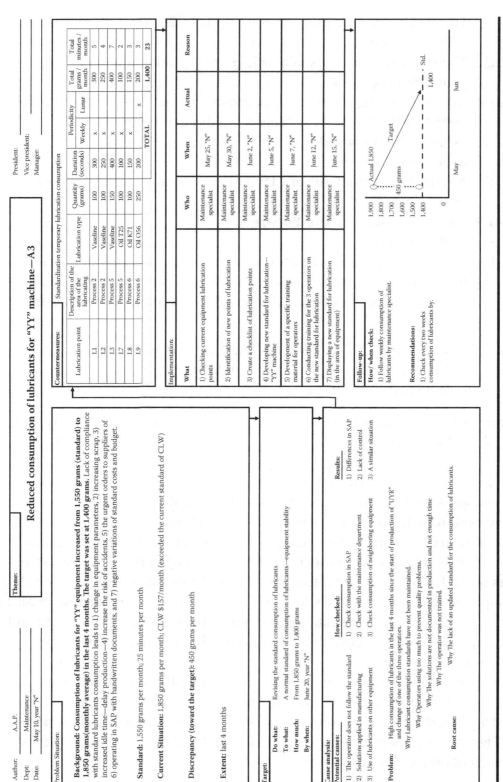

Figure 6.11 Reduced consumption of lubricants for "YY" machine with A3.

6.2.3 Conclusion

Following the MCPD system project, the annual MCI goal for PFC2 of $850,000 was achieved and the extension to PFC1 was decided. The management team at BB-Plant has been aware that the MCPD system's stake is to ensure the continuous *uncovering of hidden reserves of profitability* by continuously transforming processes based on the continuous absorption of market signals and by meeting the annual MCI targets and means.

Reference

Posteucă, A. and Sakamoto, S., 2017. *Manufacturing Cost Policy Deployment (MCPD) and Methods Design Concept (MDC): The Path to Competitiveness.* New York: Taylor & Francis.

Bibliography

Ahuja, I. P. S. and Khamba, J. S., 2008. Total productive maintenance: Literature review and directions. *International Journal of Quality & Reliability Management*, 25(7), 709–756.

American National Standards Institute, 1983. *Industrial Engineering Terminology*. Hoboken, NJ: Wiley-Interscience.

Bodek, N., 2004. *Kaikaku: The Power and Magic of Lean: A Study in Knowledge Transfer*. Vancouver, WA: PCS Inc.

Budugan, D. and Georgescu, I., 2009. Cost reduction by using budgeting via the Kaizen method. *Annals of the University "Alexandru Ioan Cuza" Iași, Romania*, 56, 3–9.

Burnham, D. C., 1972. *Productivity Improvement*. New York: Columbia University Press.

Burrows, G. and Chenhall, R. H., 2012. Target costing: First and second comings. *Accounting History Review*, 22(2), 127–142.

Carlson, C. R. and Wilmot, W. W., 2006. *Innovation: The Five Disciplines for Creating What Customers Want*. New York: Crown Business.

Chau, V. S. and Witcher, B. J., 2008. Dynamic capabilities for strategic team performance management: The case of Nissan. *Team Performance Management*, 14(3/4), 179–191.

Chiarini, A. and Vagnoni, E., 2014. World-class manufacturing by Fiat. Comparison with Toyota Production System from a Strategic Management, Management Accounting, Operations Management and Performance Measurement dimension. *International Journal of Production Research*, 53(2), 590–606.

Cooper, R., 1995. *When Lean Enterprises Collide: Competing through Confrontation*. Boston, MA: Harvard Business Press.

Cooper, R. and Slagmulder, R., 1999. Developing profitable new products with target costing. *Sloan Management Review*, 40(4), 23–33.

Cudney, E. A., 2009. *Using Hoshin Kanri to Improve the Value Stream*. Boca Raton, FL: CRC Press.

Deming, W. E., 1986. *Out of the Crisis*. Cambridge, MA: Massachusetts Institute of Technology, Center for Advanced Engineering Study.

Drucker, P. F., 2006. *The Practice of Management*. New York: HarperCollins.

Duarte, J. E., 1993. Policy deployment: Planning methods that get results. *CMA Magazine*, 67(4), 13.

Ellram, L. M., 2006. The implementation of target costing in the United States: Theory versus practice. *Journal of Supply Chain Management*, 42(1), 13–26.

Gadiesh, O. and Gilbert, J. L., 1998. Profit pools: A fresh look at strategy. *Harvard Business Review*, 76(3), 139–147, May–June 1998.

Gåsvaer, D. and von Axelson, J., 2012. Kaikaku-radical improvement in production. In *International Conference on Operations and Maintenance*. Singapore: World Academy of Science, Engineering and Technology, pp. 758–765.

Heap, J., 2008. Innovation and enterprise: The foundations of developing productivity. *International Journal of Productivity and Performance Management*, 57(6), 434–439.

Helmrich, K., 2003. *Productivity Process: Methods and Experiences of Measuring and Improving*. Stockholm, Sweden: International MTM Directorate.

Hino, S., 2006. *Inside the Mind of Toyota: Management Principles for Enduring Growth*. New York: Taylor & Francis Group.

Hutchins, D., 2008. *Hoshin Kanri: The Strategic Approach to Continuous Improvement*. Burlington, VA: Gower Publishing.

Jackson, T. L., 2006. *Hoshin Kanri for the Lean Enterprise: Developing Competitive Capabilities and Managing Profit*. New York: CRC Press.

Juran, J. M., 1995. *Managerial Breakthrough: The Classic Book on Improving Management Performance*. New York: McGraw-Hill.

Kaplan, R. S., 1990. *Measure for Manufacturing Excellence*. Boston, MA: Harvard Business School Press.

Kaplan, R. S. and Norton, D. P., 1996. *The Balanced Scorecard: Translating Strategy into Action*. Boston, MA: Harvard Business School Press.

Kondo, Y., 1998. Hoshin Kanri—A participative way of quality management in Japan. *The TQM Magazine*, 10(6), 425–431.

Lee, J. Y. and Monden, Y., 1996. Kaizen costing: Its structure and cost management functions. *Advances in Management Accounting*, 5, 27–40.

Liker, J., 2004. *The Toyota Way: 14 Management Principles from the World's Greatest Manufacturer*. New York: McGraw-Hill Education.

Maskell, B. H., 2012. *Practical Lean Accounting*. Boca Raton, FL: CRC Press.

Maskell, B. H. and Jenson, R., 2000. Lean accounting for lean manufacturers. *Manufacturing Engineering*, 125(6), 46–53.

Mather, H., 1986. *Competitive Manufacturing*. Upper Saddle River, NJ: Prentice Hall.

Mitchell, F., Nørreklit, H., and Jakobsen, M., 2013. *The Routledge Companion to Cost Management*. New York: Taylor & Francis.

Monden, Y., 1992. *Cost Management in the New Manufacturing Age: Innovations in the Japanese Automotive Industry*. New York: Productivity Press.

Monden, Y., 2000. *Japanese Cost Management*. London, U.K.: World Scientific, Imperial College Press.

Monden, Y., 2012. *Toyota Production System: An Integrated Approach to Just-in-Time*. Boca Raton, FL: CRC Press.

Monden, Y., Kosuga, M., Nagasaka, Y., Hiraoka, S., and Hoshi, N., 2007. *Japanese Management Accounting Today*. Singapore: World Scientific Publishing.

Murata, K. and Katayama, H., 2009. An evaluation of factory performance utilized KPI/KAI with Data Envelopment Analysis. *Journal of the Operations Research Society of Japan*, 52(2), 204.

Posteucă, A., 2013. Geen Lean methodology: Enterprise energy management for industrial companies. *Academy of Romanian Scientists—"Productica" Scientific Session*, 5(1), 17–30.

Posteucă, A. and Sakamoto, S., 2017. *Manufacturing Cost Policy Deployment (MCPD) and Methods Design Concept (MDC): The Path to Competitiveness*, Taylor & Francis, New York, NY.

Posteucă, A. and Zapciu, M., 2013. Quick changeover: Continuous improvement and production costs reduction for plastic-molding machines. *The Seventh International Working Conference*, Belgrade, Serbia, June 3–7, 2013, Vol. 1, pp. 141–147.

Posteucă, A. and Zapciu, M., 2015a. Beyond target costing: Manufacturing cost policy deployment for new products. *Applied Mechanics and Materials, Switzerland*, 809, 1480–1485.

Posteucă, A. and Zapciu, M., 2015b. Continuous improvement of the effectiveness of equipment driven by the dynamics of cost reduction. *Sustainable Design and Manufacturing—KES International Conference*, Seville, Spain.

Posteucă, A. and Zapciu, M., 2015c. Process innovation: Holistic scenarios to reduce total lead time. *Academy of Romanian Scientists—"Productica" Scientific Session*, 7(2), 79–88.

Posteucă, A. and Zapciu, M., 2015d. Setup time and cost reduction in conditions of low volume and overcapacity. *University "Politehnica" of Bucharest, Science Bulletin, Series D*, 77(4), 325–336.

Riggs, J. L. and Felix, G. H., 1983. *Productivity by Objectives Results-Oriented Solutions to the Productivity Puzzle*. Englewood Cliff, NJ: Prentice-Hall.

Sakamoto, S., 1992. Design concept for methods engineering. In Hodson, W. K. (Ed.), *Maynard Industrial Engineering Handbook*. New York: McGraw-Hill.

Sakamoto, S., 2006. Methods design concept: An effective approach to profitability. *Journal of Philippine Industrial Engineering*, 3(2), 1–11.

Sakamoto, S., 2009. *Return to Work Measurement*. Norcross, GA: Industrial Engineering, p. 24.

Sakamoto, S., 2010. *Beyond World-Class Productivity: Industrial Engineering Practice and Theory*. London, U.K.: Springer.

Schonberger, R. J., 1986. *World-Class Manufacturing*. New York: The Free Press.

Stamatis, D. H., 2010. *The OEE Primer: Understanding Overall Equipment Effectiveness, Reliability, and Maintainability*. Boca Raton, FL: CRC Press.

Suzuki, T., 1994. *TPM in Process Industries*. New York: Taylor & Francis Group.

Swedish Federation of Productivity Services, 1993. *SAM Training Program*. Stockholm, Sweden: Swedish Federation of Productivity Services.

Tanaka, T., 1994. Kaizen budgeting: Toyota's cost-control system under TQC. *Journal of Cost Management for the Manufacturing Industry*, 8(Fall), 56–62.

Taylor, F. W., 1911. *The Principles of Scientific Management*. New York: Harper & Brothers.

Tennant, C. and Roberts, P., 2001. Hoshin Kanri: A tool for strategic policy deployment. *Knowledge and Process Management*, 8(4), 262–269.

Tuttle, T. C., 2016. *Growing Jobs: Transforming the Way We Approach Economic Development*. Santa Barbara, CA: ABC-CLIO, LLC.

Witcher, B. J., 2003. Policy management of strategy (Hoshin Kanri). *Strategic Change*, 12(2), 83–94.

Witcher, B. J. and Chau, V. S., 2007. Balanced scorecard and Hoshin Kanri: Dynamic capabilities for managing strategic fit. *Management Decision*, 45(3), 518–538.

Womack, J. P. and Jones, D. T., 1996. *Lean Thinking: Banish Waste and Create Wealth in Your Corporation*. New York: Free Press.

Womack, J. P., Jones, D. T., and Roos, D., 1990. *The Machine That Changed the World: The Story of Lean Production—Toyota's Secret Weapon in the Global Car Wars That Is Now Revolutionizing World Industry*. New York: Free Press.

Yamamoto, Y., 2013. Kaikaku in production toward creating unique production systems. PhD thesis, Department of Innovation, Design and Engineering, Mälardalen University, Eskilstuna, Sweden.

Yamashina, H. and Kubo, T., 2002. Manufacturing cost deployment. *International Journal of Production Research*, 40(16), 4077–4091.

Zammori, F., Braglia, M., and Frosolini, M., 2011. Stochastic overall equipment effectiveness. *International Journal of Production Research*, 49(21), 6469–6490.

Zandin, K. B., 1980. *MOST Work Measurement System*. New York: Marcel Dekker.

Zanoni, S., Ferretti, I., and Tang, O., 2006. Cost performance and bullwhip effect in a hybrid manufacturing and remanufacturing system with different control policies. *International Journal of Production Research*, 44(18–19), 3847–3862.

Zawawi, N. H. M. and Hoque, Z., 2010. Research in management accounting innovations: An overview of its recent development. *Qualitative Research in Accounting & Management*, 7(4), 505–568.

Zengin, Y. and Ada, E., 2010. Cost management through product design: Target costing approach. *International Journal of Production Research*, 48(19), 5593–5611.

Zoysa, A. D. and Herath, S. K., 2007. Standard costing in Japanese firms: Reexamination of its significance in the new manufacturing environment. *Industrial Management & Data Systems*, 107(2), 271–283.

Index